A FIELD PHILOSOPHER'S GUIDE TO FRACKING

Adam Briggle

LIVERIGHT PUBLISHING CORPORATION

A DIVISION OF W. W. NORTON & COMPANY

NEW YORK LONDON

A FIELD PHILOSOPHER'S GUIDE TO FRACKING

How One Texas Town
Stood Up to Big Oil and Gas

For information about permission to reproduce selections from this book,
write to Permissions, Liveright Publishing Corporation,
a division of W. W. Norton & Company, Inc.,
500 Fifth Avenue, New York, NY 10110

For information about special discounts for bulk purchases, please contact
W. W. Norton Special Sales at specialsales@wwnorton.com or 800-233-4830

Manufacturing by RR Donnelley, Harrisonburg, VA
Book design by Lovedog Studio
Production manager: Anna Oler

Library of Congress Cataloging-in-Publication Data

Briggle, Adam.
A field philosopher's guide to fracking : how one Texas town stood up
to big oil and gas / Adam Briggle. — First edition.
pages cm
Includes bibliographical references and index.
ISBN 978-1-63149-007-1 (hardcover)
1. Hydraulic fracturing—Political aspects—Texas. 2. Gas wells—
Hydraulic fracturing—Political aspects—Texas. 3. Oil wells—Hydraulic
fracturing—Political aspects—Texas. 4. Denton (Tex.) I. Title.
QE431.6.M4B75 2015
363.11'962233809764555—dc23
2015020528

Liveright Publishing Corporation
500 Fifth Avenue, New York, N.Y. 10110
www.wwnorton.com

W. W. Norton & Company Ltd.
Castle House, 75/76 Wells Street, London W1T 3QT

1 2 3 4 5 6 7 8 9 0

For MG and Lulu

Daß ich erkenne was die Welt
Im Innersten zusammenhält . . .

—Goethe, *Faust*

My dear Phaedrus, whence come you, and whither
are you going?

—Plato, *Phaedrus*

CONTENTS

PART III. SEE NO EVIL

PART IV. RESPONSIBLE DRILLING

Photographs to follow page 176.

MAPS AND FIGURES

Maps

Figures

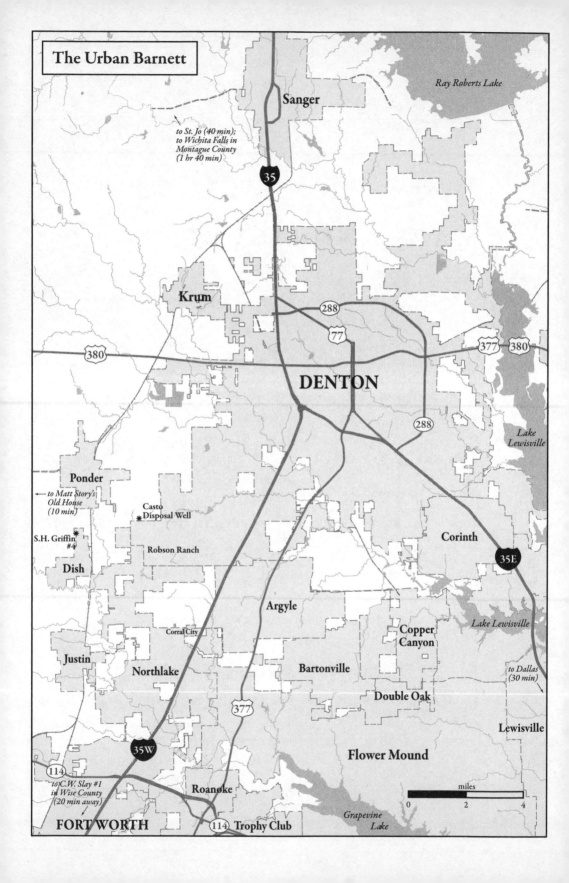

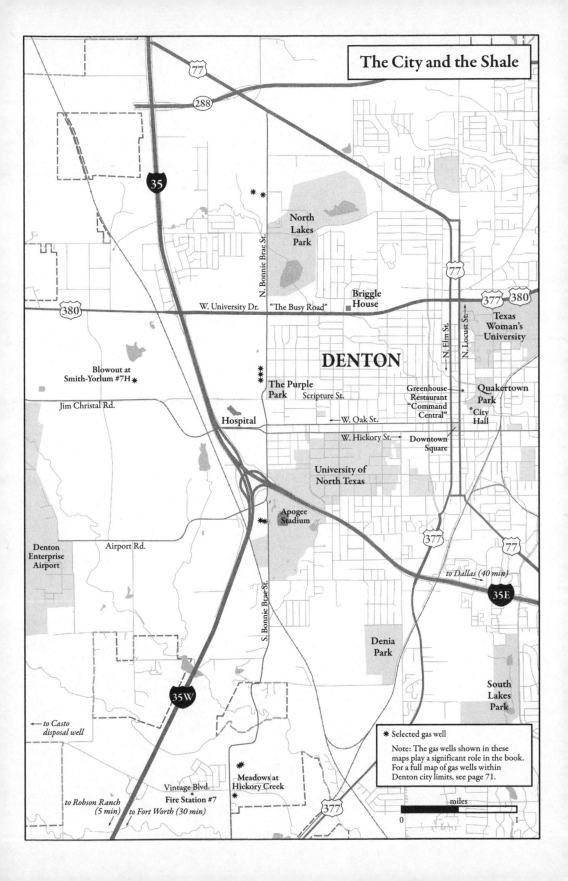

A FIELD PHILOSOPHER'S GUIDE
TO FRACKING

LET A THOUSAND GAS WELLS BLOOM

IHAD NEVER HEARD OF FRACKING UNTIL ONE FINE autumn day at McKenna Park in Denton, Texas. I was pushing my son, MG, on the swing. Just down the hill and across the street was Rayzor Ranch, marking the western edge of development. As MG went up in the air I could see it stretch for nearly a mile toward the setting sun below his feet—a hardscrabble of dusty brush and rugged prairie grass. Like so much of the sprawling Dallas–Fort Worth metroplex, Rayzor Ranch was slated to be transformed from mesquite and cattle into Walmart and Petco.

A young mother pushing her little boy in the swing next to us said, "They're going to drill three more gas wells there. Won't be long until they start fracking." She said it so matter-of-factly that I couldn't bring myself to ask what she was talking about. I just stood there staring. The only sound was the rhythmic squeaking of the chains on the swings as the kids went up and down.

It was 2009, and we had just moved to Denton so that I could begin my new job as a philosophy professor at the University of North Texas (UNT). I didn't know it then, but that day at the park would set me on a quest to develop a philosophy of fracking. I started to educate myself and learned there were nearly 15,000 gas wells on the Barnett Shale at that time. Around 250 of them were inside the city limits of Denton. I learned that fracking had sparked a global energy revolution and that it all began just a few miles west of our new hometown.

About a year later, with the release of the documentary *Gasland* and widespread drilling on the Marcellus Shale in Pennsylvania, fracking began to enter the collective consciousness of the nation. And urban drilling became the most contentious political issue in Denton and many other towns and cities atop shale plays.

That's when Kevin Roden, a Denton City Council member, approached the Center for the Study of Interdisciplinarity (CSID) at UNT. I was a faculty fellow at CSID then, and Kevin asked me if I would form a group of citizens to advise the city on its gas well regulations. Kevin was a PhD student in philosophy and shared my sense that this issue cut across the disciplines and was driven by deeply rooted values and worldviews. I realized this was my chance to practice what the members of CSID had been calling "field philosophy." So I took Kevin up on his offer and started helping people think through the ethics questions that arise when drilling rigs go up next to neighborhoods and playgrounds.

Like Socrates, field philosophers wander around questioning purported experts, trying to figure out what we really know and which choices we should make. Rather than talk only to other academic philosophers, I talked to scientists, engineers, policymakers, citizens, and members of the industry. I also blogged, organized educational events, gave presentations at community rec centers, produced videos for You-Tube, toured frack sites, spoke with concerned parents, wrote op-eds, and even had an uncanny visit from the FBI.

Eventually, after years of failed efforts to control the industry, I worked with several amazing people against impossible odds to make Denton the first city in Texas to ban hydraulic fracturing. Along the way, I was accused of being a radical and a Russian. I made some powerful enemies. But I also made many friends, and I believe that together we made Denton a better place. And, perhaps most importantly, we became a voice in the national and even international conversation about fracking and the ethics of energy. We became a ray of hope for grassroots democracy in an age of big money.

For the first couple of years of my involvement, the more I studied fracking, the more dissatisfied I became with the growing chatter about it. The fracking debate is scattered across a hyperpaced and balkanized media landscape, squeezed into simplifications about truth and lies, and partitioned into interest-group politics. The effect was disorienting.

For a long time, I struggled to make sense of it all. Surely fracking tells us something important about the way we live. I just needed to pull the strands together into a coherent story that explains and illuminates our world. Mineral rights, polluted water, extreme energy, campaign contributions, citizen scientists, psy-ops techniques, local bans, dueling experts . . . how does it all fit together? As I read, listened, and blogged, slowly a pattern emerged about fracking and the character of the modern world.

Fracking exemplifies the technological wager, by which I mean a gamble or even a faith that we can transform the world in the pursuit of narrowly defined goals and successfully manage the broader unintended consequences that result.[1] In many ways, we are gambling on the success of future innovations to bail us out of problems created by present innovations. I think that if we are to live with high technology we cannot avoid this wager. The question is whether we can establish conditions to make it a fair and reasonable bet. In the case of fracking, I will argue, these conditions are largely not in place.

The term "fracking" is often used as shorthand for "hydraulic fracturing," a well-stimulation technique that blasts rock formations with sand, water, and chemicals in order to extract oil and gas. The ban in

Denton applied only to hydraulic fracturing understood in this sense, but to always confine ourselves to such a narrow frame of reference is to lose sight of the bigger picture. That's why in this book I will use the term more broadly to refer to the intensification of hydraulic fracturing and its convergence with other innovations like 3-D seismic imaging and horizontal drilling. Beginning roughly at the turn of the twenty-first century, fracking made oil and gas production from unconventional reserves like the Barnett Shale in Texas and the Marcellus Shale in Pennsylvania economically viable.

As an example of the technological wager, fracking is a window into the ethics and politics of the human-built world. It is a story about the character of engineering and the way in which innovation happens, especially in the context of American neoliberalism.

There is certainly something praiseworthy about a profession dedicated to improving the lot of humanity by conquering nature. But there is also ambiguity, a kind of moral murkiness, forged in the core of engineering. After all, engineers' creations both liberate and enslave, destroy and uplift. And despite all the systematic calculation that goes into their work, the scope of their impacts always outstrips the scope of their thought.

Like the sorcerer's apprentice, engineers unleash events beyond their capacity to understand or control. By working with models that inevitably simplify the world, they exercise power over reality by overlooking most of its complexity. Their goal is functional adequacy, not mirroring the world.[2] This means that engineers often overlook important things. Consider a few simple examples. Asbestos was designed for insulation. Chlorofluorocarbons (CFCs) were engineered for use as refrigerants and propellants. Tetraethyl lead (TEL) was added to gasoline to boost engine performance and fuel economy. They all performed their intended functions admirably. Of course, they also caused the unintended effects of cancer, ozone depletion, and lead poisoning. Those effects were overlooked. They weren't accounted for in the models.

The basic point about changing the world by overlooking its complexity holds more generally. To get microwave ovens to work, no one needs to account for their impacts on the quality of food and family mealtime. Automobiles can be made to transport people without modeling effects on the climate, the character of cities (think sprawl), or the geopolitics of oil. Computers can process data independent of any thought about information overload, cyber terrorism, or e-waste. Industrial fertilizers can grow corn absent any thought about algal blooms in the Gulf of Mexico. Pesticides can be made to work without modeling impacts on monarch butterflies. You can make a perfectly good carpet without asking what happens to it after its useful life.[3] In the case of fracking, you can get the fossil fuels out of the ground at a profit without accounting for the long-term fate of chemicals, water and air quality, climate impacts, seismicity, or community livability.

In all these cases, we face the same question: What should we do about the fact that freedom to pursue intended desired outcomes also creates unintentional harms?

One response is to ask engineers and their corporate employers to take more factors into account prior to enrolling their technologies into society. This is usually couched in terms of the precautionary principle, which sees the technology as guilty until proven innocent. Those who adopt this standpoint focus, for example, on all the harm that could have been avoided had we just complexified our up-front assessment of, say, asbestos and asked about its potential to cause cancer rather than just about its potential to insulate pipes. The state of New York took a precautionary approach to fracking. It implemented a six-year moratorium while conducting studies about potential health, environmental, and economic impacts. Just six weeks after we passed a ban in Denton, New York governor Andrew Cuomo announced that hydraulic fracturing would also be prohibited in the state of New York. The state health commissioner, Dr. Howard Zucker, said their studies found "significant public health risks" and insufficient evidence to establish that hydraulic fracturing could be done safely. "The potential risks are too great," he said. "In fact, they are not even fully known."[4]

But what does it mean to "fully know" something? There is no way to prove a technology is innocent. In the early stages of development, it is impossible to predict how it will play out. To demand certainty about broader impacts is to postpone action indefinitely. So if we are going to get the intended benefits of a technology, we must also risk getting unintended harms. At some point, we must roll the dice and move the experiment out of the lab and into the world.[5]

That's why it is important to consider not only how much we know about a technology prior to committing to it. We also need to think about the structure of these real-world experiments. We often think of experiments as occurring in the laboratory, where conditions can be tightly controlled. What I have in mind is something closer to field experiments, where, say, researchers examine the effectiveness of a poverty-reduction policy as it plays out in messy reality. We intervene in the world, hoping our intentions will pan out but also aware that unintended things might happen.[6] We can call it an experiment, but it's good to keep in mind that all experiments are a kind of gamble— we are betting on an idea, and the stakes of the wager are highest when it unfolds in real life. What conditions must be in place to make real-world experiments reasonable and fair? I think there are three:

1. Those most vulnerable to the unintended harms must give their consent to the risk, or, at the very least, they must be compensated for any harms incurred.
2. There must be a robust system for monitoring and learning from the real-world experiment.

3. The experiment must be modifiable when problems are identified through the learning process; that is, the original innovation must be readily renovated.

Fracking often fails all three conditions and is therefore largely
an unjust real-world experiment. At least, that is my central claim. I
developed this idea primarily through my experiences in Denton and
talking to people most affected by fracking—people you will meet
in this book, like Cathy, Maile, and Calvin. Many of the things that
went wrong in Denton often, but not always, go wrong elsewhere.[7]
They provide important lessons for how to run more ethical real-world
experiments.

I have a precautionary disposition; I don't easily give up a known
good for an unknown better. But I am not making a precautionary argument in this book. Rather, I am entering sympathetically
into a *proactionary* worldview. The transhumanist philosopher Max
More first sketched the "proactionary principle"—his antidote to
precaution—in triumphant terms:

Let a thousand flowers bloom! By all means, inspect the flowers
for signs of infestation and weed as necessary. But don't cut off the
hands of those who spread the seeds of the future.[8]

It is an alluring worldview. It says that rather than avoid error we
should take risks in the pursuit of profound truths and great rewards.[9]

With the fracking revolution, the United States has certainly let a
thousand gas wells bloom. Actually, from 2001 to 2011, an additional
170,000 gas wells were drilled, and between 2008 and 2013 domestic
crude oil production increased 50 percent.[10] As this boom comes to
back yards, it has caused concerns about its environmental and health
impacts. Yet little is known about those impacts. One scientific study
called it "an uncontrolled public health experiment on an enormous

scale."[11] A major review of the literature concluded that several "data gaps" persist even after fifteen years of intensive fracking activity across the country.[12]

The oil and gas industry denies this. After all, no one wants to be accused of experimenting with people's lives. The industry is fond of saying that hydraulic fracturing has been around for sixty years. It's tried and true, they claim. In one sense they are right: Halliburton (a major service provider in the oil and gas business) was fracking wells in the early 1950s. But back then they were using an average of 750 gallons of fluid and 400 pounds of sand per well.[13] Today's oil and gas wells can average between 5 and 8 million gallons of fluid and 4 million pounds of sand. That's a 10,000-fold increase. And this is not to mention the huge growth in the number of chemicals used (now over 750), the power of the pumps used to pressurize the concoction and crack the shale, and the convergence of these techniques with horizontal drilling and seismic imaging.

Comparing the fracked wells of the 1950s with those of today is like comparing a stick of dynamite, which was first patented in 1867, with a modern bunker-buster bomb. Sure, they are a continuation of sorts, but the differences are more important than the similarities. Things are souped up to a whole new extreme.

Furthermore, if fracking is such an established technology, what explains the massive production increases at the turn of the twenty-first century? And why was it only in the last few years that astronauts looking down on Earth at night could see the Bakken Shale of North Dakota and the Eagle Ford Shale of Texas lit up like giant cities of gas flares? And why, even as late as 2010, was the CEO of Schlumberger, the world's largest oil-field services company, saying, "I don't think that the actual optimum technology set for producing shale gas has yet been defined—at the moment, we are doing it by brute force and ignorance"?[14] It's because contemporary fracking is a new kind of creature. And we don't know what it is going to do to our water, our air, and our climate. When it comes to the impacts of fracking, as one policy analyst said, "the science is all over the place." [15]

It really is an experiment.[16] The intriguing thing about Max More's idea of proaction is that real-world experiments are not necessarily bad. Indeed, they are unavoidable, that is, if we want to continue lives of ever more consumption of ever newer things. The question is not whether we are conducting real-world experiments; the question is, how do we make them fair?

That's why the proactionary principle is a call to embrace risk, not recklessness. Tempering even More's techno-enthusiasm is the ethic of innovation I noted above with its three criteria: we must seek the consent of, or at least compensate, the vulnerable, we must "inspect" (monitor) the real-world experiment, and we must "weed" (renovate) it when problems arise. Without those conditions any "principled" alternative to precaution is no more than rapacious corporate capitalism. And that is what the story of fracking makes all too clear.

Rex Tillerson, the CEO of ExxonMobil, argued it would be "very dangerous" to turn the tables and make oil and gas companies prove safety prior to action. "If you want to live by the precautionary principle," he said, "then crawl up in a ball and live in a cave." [17] Yet even if we accept his jaded view of paralysis by precaution, we don't need to abandon all principles and settle for the unbridled pursuit of profits. There is a proactionary alternative both to strict precaution and to ruthless market utilitarianism. But, as the fracking case shows, living up to this alternative ethical standard requires us to rethink much about the politics of technology.

WHEN I WAS an undergraduate, I wanted to pursue a PhD in philosophy. So, one day, I asked a philosophy professor about this. He was the stereotypical academic with an unruly explosion of black hair and thick, round-rimmed tortoiseshell glasses. He just said, "Come with me." And he skittered to his office, where he pulled a book off one of his shelves and handed it to me. It was so full of jargon that I couldn't make heads or tails of it. "If this is what you want to do, then by all means become a philosopher." He grabbed the book and tossed it on the table, where it landed with a whap and a poof of dust. I got the message: if you choose philosophy you'll be writing technical treatises destined to collect dust. I wanted to be involved in real-world problems, so I decided to pursue a PhD in environmental studies instead.

Philosophy wasn't always so inward-looking. That only began about a hundred years ago when it started to develop as a professional academic discipline. Back then, the pragmatist philosopher John Dewey lambasted the young profession for becoming irrelevant, "sidetracked from the main currents of contemporary life." He argued that "philosophy recovers itself when it ceases to be a device for dealing with the problems of philosophers and becomes a method, cultivated by philosophers, for dealing with the problems of men." [18] That's a pretty good motto for the field philosopher.

Despite Dewey's admonitions, though, philosophy in the twentieth century largely became a technical exercise of no particular interest to anyone save a small cohort of disciplinary peers who seemed to take societal irrelevance as a sign of intellectual seriousness. Just as our growing technoscientific powers were unleashing profoundly philosophical questions about justice, our proper place in the natural world, and the very meaning of being human, philosophers retreated to the sidelines.

Of course, a gap has always existed between the concerns of philosophers and our real-world philosophic problems. We all struggle with how best to live—how to raise our children, how to prioritize our time, how to handle conflict, and on and on. But when these questions are formalized in dialogues, they tend to fossilize and start to look like navel gazing. Now, figuring out how to live a good life is not irrelevant. In fact, it's the most relevant thing for a mortal with some degree of free will to do. The problem is that constantly questioning assumptions is not a very practical way to get along in the world.

The tension between philosophers and society has been with us since the time of Socrates. With one foot in contemplation and the other in contemporary affairs, philosophy has alternately bored and irritated the outside world. Indeed, philosophers themselves have long complained about the uselessness of philosophy. Descartes scorned the abstractions of the Schoolmen, and Marx said the point of philosophy was to change, rather than merely interpret, the world.

In the twentieth century, the gap between philosophy and society grew into a chasm.[19] The research university disciplined philosophers, placing them in departments, where they wrote for and were judged by their disciplinary peers. Philosophers are often ridiculed for unsuccessfully aping the *cognitive* form of the sciences—they try to be scientists but never produce results. But the real problem is their all too successful aping of the *institutional* form of the sciences—they have become insular specialists just like every other discipline.

This is starting to change. A movement is afoot in philosophy. It's part of a broader movement for academics to be more socially engaged in the face of increasing calls for intellectuals to demonstrate the relevance and impact of their work. In philosophy, this trend started with bioethics, which puts philosophers in conversations with stakeholders struggling with issues from personal questions about end-of-life care to political questions about cloning, stem cell research, healthcare, and biomedical enhancements. By now the movement is diverse, marching under several banners, including public philosophy, philosophy and the city, experimental philosophy, feminism, field philosophy, and more. I welcome all of these innovations.

The term "field philosophy" is the brainchild of Bob Frodeman, my mentor and advisor while I was a student getting my PhD in environmental studies at the University of Colorado at Boulder. He is now

my colleague and friend in the Philosophy and Religion Department
at UNT. Actually, when I met him, he was talking about "humanities
policy," saying the humanities can be just as useful for policymaking
as the social and natural sciences. He had experienced this firsthand
working with the US Geological Survey to help them think through
the ethical and metaphysical questions posed by their work.[20]

Bob has always been the black sheep of his discipline and even got
kicked out of one philosophy program as a student for "not being a real
philosopher." He went on to study at Penn State under the guidance
of Alphonso Lingis, a wild soul who left his cockatoos at home every
summer to haul two suitcases crammed with the works of Plato and
Heidegger to some incredibly dangerous part of the world like Beirut
to think deep thoughts amidst the chaos of life. With Professor Lingis
for a mentor, it's no surprise that as part of his dissertation Bob built a
cabin in the woods.

When we first met in 2003 in Boulder, Bob teased me about my
long hockey hair and baseball cap (I have since cut my hair, in part
to survive the Texas summers). But it didn't take long for us to see we
were kindred spirits, in search of a different future for philosophy.

When Bob and I met in Boulder he was working at the Center for
Science and Technology Policy Research. He had used his slick talking
skills—gleaned in equal parts from Aristotle's *Rhetoric* and his salesman father—to convince Roger Pielke Jr., the center's director, that
policy questions are often philosophical questions in disguise. Public
policies about, say, fracking, are just the boats riding atop waves driven
by deeper cultural and intellectual currents. You can't understand
where the boats are going if you don't understand the ocean.

We tend to think that our problems are basically technical and scientific in nature. And our culture has little patience for philosophical
questioning, especially if it threatens to slow the rate of innovation.
Yet although everyone appeals to science to resolve political disputes

with technical dimensions, they just turn around and cherry-pick the science that supports their foregone conclusion.

Science is obviously crucial, but it alone cannot answer questions, for example, about whether a proposed fracking project is safe enough. That depends on how one perceives and is situated in relation to the risks; everything is framed by, and interpreted through, values and worldviews. That's why I have always thought the most important question with fracking was the jurisdictional one—not *what* should the rules be, but rather *who* should make the rules.

I'm not saying everything is subjective. I'm trying to get us to see past our culture's objective/subjective dualism. We tend to put propositions into one of two boxes: either they mirror the world as it really is, or they are merely opinions, desires, or wishful thinking. The fracking debate is a prime example, because both sides frame it in black-and-white terms of brute facts vs. myths (plain truths vs. skullduggery; reason vs. emotion; etc.). The result is a secular rehash of the medieval *odium theologicum* when religious factions bashed each other over the head with incommensurable claims to absolute truth.

Now, sometimes things are black and white. But usually they are gray. Even as my opposition to fracking intensified, I always recognized that you could paint a very rosy picture of it without resorting to any lies whatsoever—you just have to frame it in terms of jobs, economic growth, lower prices, energy security, and displaced combustion of dirty coal. Of course, that framing boxes out other dimensions that occupied my mental field of vision—air and water pollution, intensified fossil fuel reliance in an age of climate change, earthquakes, disenfranchisement of communities, and neighborhood industrialization. The tension here can't be reduced to objective facts vs. subjective opinions. This isn't about what the world is really like. It's about what's the most appropriate way to see things given a plurality of legitimate views.

It is exhausting to see things in the round—to continuously hold alternative views on the world as live options. It forces you to constantly transport yourself outside of your assumptions to get a glimpse from another angle. It is so tempting to lock into one framing or the other with certainty and finality. It's the seduction of simplicity—to convince yourself that you have arrived at the end of the arduous trail that is thinking. But you would only be fooling yourself. That's what Socrates showed time and again when he stumped people who were oh so sure of themselves.

And that's why field philosophers are vital today. With its culture of disciplinary expertise, our age is so very sure of itself. Yet the artificial silos of our specializations break down in the face of the wicked problems of the real world. We pretend we can carve up reality into discrete boxes of knowledge, but things are always a messy hybrid drenched in uncertainty and politics. The field philosopher is not an expert but a second-order expert—watching the way knowledge moves through networks, is constructed and contested. Things are so complex that everyone's knowledge (even the expert's) quickly reaches its end far before the problem is defined, let alone resolved.

Perhaps the central remit of the field philosopher is to show how we cannot escape thinking. We cannot help but live in a world of rich hermeneutic possibilities. We cannot delegate thinking to the experts, because even they do not know. We are doomed to philosophize.

One basic problem the field philosopher is working on is that people make unavoidable value judgments through muddled and unexamined assumptions. Or, worse, experts (a category various agencies and industries try to lay claim to) pass value judgments off as if they were simply "the facts" as churned out by a neutral method. The role of the field philosopher is to excavate and examine the ethical, aesthetic, and even metaphysical presumptions that are inevitably packed into the black box of expert discourse and political messaging. Hopefully, this clarifies the debate and helps people weigh the merits of competing frames and arguments. The idea is to make real-time contributions, rather than write something up later to be published in a peer-reviewed journal that no one involved with the policy will ever read.

As I got pulled deeper into the politics of fracking, I became intimately aware of the danger that lurks in the heart of any attempt to conduct socially engaged scholarship: you might purchase social relevance at the cost of your philosophical soul. This is a hazard that the philosopher Raphael pointed out in Thomas More's 1516 book *Utopia*. Raphael had been to the island of Utopia and had gleaned wisdom there that could benefit the people of Europe. But he knew the kings would not listen, because his recommendations would require rethinking basic assumptions (for example, he recommended abolishing private property). Policymakers have no time for philosophy, he said; they are not open to questioning established values even when reason makes a compelling case that they ought to do so.

Raphael chose to remain silent and distant from political affairs, like a desert father wise in his seclusion. He believed that philosophers can question entrenched conventions and habits of thought only among themselves and on the margins of society. If you try to counsel and serve society, you will be shoehorned into a practicality defined by the very standards and assumptions any good philosopher would question. To put it into more contemporary terms, if you are working with the people, then you'll be judged by their criteria. A popularity contest will displace the pursuit of wisdom: the best philosopher becomes the one with the most Likes on Facebook.

This gets to important questions about democracy, because field philosophers are most certainly not philosopher kings—they work within democratic institutions, such as they may be in our capitalist society. Yet our commitment to liberal democracy arguably means that we are also committed to the notion that all adults (save, perhaps, the insane) are qualified to participate as citizens. Since practicing citizenship hinges on making various philosophical judgments (about, for example, what constitutes the right thing to do), this means we are committed to the idea that all adults are basically philosophers. The role of the field philosopher, then, is to assist people in doing the philosophical thinking they are already implicitly doing.

But how much do we actually think openly and deeply about the choices we collectively face? The philosopher's main tool is rational persuasion, but does that have any oomph in our society? On a skeptical reading, at least, it doesn't. People have their minds made up (they persist in their beliefs even in the face of contrary evidence and expert claims).[21] Politicians lie, activists simplify, media and corporations manipulate, people are driven by fear and unconscious desires, etc. The forces of irrationalism are arrayed like a phalanx of Spartan soldiers, stretched as far as the eye can see along the horizon. Socrates will be crushed. His body will hardly register as a bump under the shields and boots a-marching.

Maybe that's too harsh. Maybe it's an unfair portrayal of human nature. But even stated more gently: we all have basic convictions that drive our political disagreements. What would it take to get you to change your mind . . . a conversation with a philosopher? Really?

Here's a little illustration of a field philosopher's dilemma. I was once reprimanded by an activist for publishing an article that was ambiguous about whether municipal fracking bans are good or bad. He called me on the phone when I was watching MG run around on the courthouse lawn in the center of Denton's downtown square. We had just got ice cream from Beth Marie's, and it was time to burn off the sugar. The self-described activist told me never to air shades of gray in public: "We have to hammer home a simple message over and over again: fracking is bad!" I told him this strategy of certainty-mongering and oversimplification was the exact same one used by the industry. "Yes," he admitted, "we have to fight fire with fire. But the difference is that we have God on our side."

The philosopher must never become a fundamentalist. Now, this doesn't mean he or she can't form strong beliefs and fight for them. Still, though, there is a tension here. Once you are advocating for a cause, doubt and questioning, arguably the philosopher's bread and butter, become liabilities. Once you are pulling with a group, you risk being praised not for sound arguments but for simplistic jibes that demonize the enemy and energize your base. Rather than opening up the ambiguities and uncertainties of any given position, you might find yourself boiling things down to quips, tweets, and memes. Rather than asking the infuriating questions, you'll start to pretend you are the expert so that you can give the people what they want: answers, which they already know anyway, and which are really an excuse to stop thinking. Before you know it, you've become an activist. Or what the ancients called a sophist. You are canvassing the streets with marching orders to keep things simple, never hint at the other side's arguments, avoid doubt, and enlist more foot soldiers.

It's a pickle. Socrates described himself as a gadfly, spurring a lazy society toward self-awareness. But some of his friends gave him arguably a more apt title, the stingray, needling everyone with a relentless barrage of questions until paralysis sets in and all action ceases. If philosophers only question ends and never guide us toward the better ends, then how can they be anything but useless? Yet if they commit themselves to a certain goal and act with others to realize it, have they stopped philosophizing?

I believe that field philosophers can legitimately form moral convictions and advocate for policies that reflect those convictions. That's what I have done, at least . . . and, really, Socrates did plenty of that, too, as have so many philosophers over the ages. But in my case, as the stakes got bigger, what had been a conversation about fracking turned into a war. I found myself in the middle of the fight. As time went on, I believed more and more strongly in our cause. But I always wondered if the battle still constituted philosophy or if philosophy had already been claimed as a casualty.

A Field Philosopher's Guide to Fracking is arranged into four parts. The title of part I, "Thales Falls into a Gas Well," is an allusion to the famous tale of how the philosopher Thales fell into a water well because he was looking up in contemplation of the stars rather than minding his earth-bound step. The first two chapters tell how, through the formation of the Denton Stakeholder Drilling Advisory Group (DAG), my story interwove with the story of Denton and the birth of fracking. In other words, they recount how it is that I came to fall into the politics of fracking. Chapters 3 and 4 continue this narrative, while establishing the conceptual framework around the key ideas of precaution and proaction.

The next three parts of the book each focus on one of the three conditions for ethical real-world experiments. Part II, "Guinea Pigs of the Shale," delves into the idea of informed consent as DAG shifts from advising City Council to launching Frack Free Denton, a citizens' initiative to ban hydraulic fracturing. Part III, "See No Evil," takes up the failures of air and water monitoring as Frack Free Denton struggles to develop into a genuine social movement. Part IV, "Responsible Drilling," looks at the challenges of renovation as it recounts the frenzied final months leading up to the improbable and inspiring vote on a rainy day in November.

Part I

THALES
FALLS INTO
A GAS WELL

CHAPTER 1

THE CITY
AND THE SHALE

I N 1919, MY GREAT-GRANDFATHER DR. DON DEAL took the train from Springfield, Illinois, to Lafayette, Louisiana. A surgeon who had invented a new appendectomy procedure, he was looking for a backwater place to get away and raise hell with his friends. Family legend, mostly gleaned from my aunt Judy, tells that Dr. Don had a mistress who would join him on his escapades to Cajun country. My dad, a shy bookworm everyone knew as Little Bobby, was younger and more naïve than his sister Judy. He only remembers that his grandpa used to give him bright pennies that he had shined at Maldaner's restaurant in downtown Springfield off of Route 66.

Dr. Don bought about a thousand acres of bayou at the sales that happened when folks couldn't pay their tax bills. His new property was on the right bank of the Atchafalaya River. This was long before the Army Corps of Engineers built the Old River Control Structure to keep the Mississippi from jumping banks and changing course to follow the Atchafalaya to the Gulf. Had that happened, New Orleans would have dried up, and much of Dr. Don's investment would have been at the bottom of the new Mississippi. Today, the property he bought is mostly controlled by Army Corps flood easements. It's carpeted in cypress trees, Spanish moss, and banana spiders, and along its edge the massive levee holds back the Atchafalaya, looming overhead like the sword of Damocles.

There's another family legend that Dr. Don was hoodwinked by a shady salesman who lit swamp gas into balls of fire to prove the land was rich with oil. But I never put much stock in that story. He probably thought he'd make some revenue from timber, but oil addiction had yet to take hold of America. At the time there was no highway system and hardly any roads in that part of the country. Henry Ford had just rolled out the Model T in 1908, and aviation was in its infancy—the first transatlantic flight occurred in 1919, the same year Dr. Don bought his land. He likely didn't think that oil would be a good reason to buy a swamp.

But with the growth of the auto industry, his southern hideaway was about to become a windfall. In the 1940s, just as the wartime boom increased oil demand and companies like Chevron ramped up their exploration activities in the bayou, Dr. Don began leasing his swampland for oil and gas development. Louisiana crude first flowed up what would be known as Don Deal Well #1 around Christmastime 1952, just a few months after Dr. Don died.

He never saw his well in production, but his wife did. My dad and his sister Judy knew her as "Gram Deal," a worldly woman who hosted young students from Europe and enjoyed a glass of Johnnie Walker Red while she watched the news on her newfangled color television. It was from her that my dad got his love of art and the St. Louis Cardinals. Gram Deal kept score of all the Cardinal games as she listened to Dizzy Dean and Harry Caray on the radio. She was already quite wealthy, but she was stunned early in 1953 to discover that her bank account had an additional $100,000 as the result of the Don Deal well. Yet even as her Louisiana assets ballooned over the following years, she still continued to garden and can her own vegetables—an unshakable habit from a childhood spent on a farm before refrigeration.

From my great-aunt Alice to her little nephew Bobby, the next two generations of the Deal family sank about half a dozen more wells on Dr. Don's land. They also profited from oil wells on neighboring property that the industry had aggregated into the Krotz Springs Unit, named for a nearby town comprised of little more than mobile homes and a country store. Some of my fondest childhood memories are from our annual visits to the family land in Louisiana, where we would bounce along in pickup trucks to see the oil wells tucked among the overcup oaks. We'd eat crawfish along the banks of the Two O'Clock Bayou as we listened to tall tales from Reggie, the crazy coonass alligator hunter. I also remember being extremely bored, spinning around in big chairs in the offices of oil companies as men in suits pointed at charts and maps.

I would later learn that oil revenues financed half of my childhood existence. They helped build our sprawling home in the foothills of Colorado Springs next to one of the golf courses owned by the five-star Broadmoor Hotel. My house was the envy of all the kids, because it had a brass fireman's pole that you could take from the living room two stories down to the basement. As I slid down that pole and played video games with my friends on our Atari, I didn't think at all about the folks in Krotz Springs, a thousand miles away, in their modest homes looking out on the Deal family oil wells and the local refinery.

In the 1970s, the price of oil spiked. This was good news for the Deal family wells. But at the same time, the production from our wells slowly started to decline, mirroring a nationwide dip in domestic oil production that would last through the rest of the century. Now, there is a strong case to be made that what the price spike indicated was not a shortage of oil but rather a weak US dollar.[1] But in the '70s many people took the increase in price as a sign of scarcity. It seemed to confirm the view that there is a finite stock of natural resources that we are rapidly depleting. At least, that's how President Jimmy Carter saw it in April of 1977, a few weeks after I was born, when he went on television to tell the nation it was "simply running out" of oil and natural gas.

Carter framed the problem as one of character: "We must not be selfish . . . [and we must] put up with inconveniences and make sacrifices."[2] It's almost as if he saw us going back to the days of Gram Deal canning vegetables on the farm after America's brief obsession with fossil-fueled luxury. He was tapping into the apocalyptic zeitgeist of the time: "silent spring," "quiet crisis," "tragedy of the commons," and "population bomb." This sense that humanity was pressing up against the limits of nature deeply impressed itself on me. As I watched Colorado Springs sprawl along the foothills, chewing up prairie and ponderosa pines and occasionally smudging Pikes Peak with a yellow film of smog, I felt an almost panicked sense of angst. Indeed, though I never called it such when I was a kid, my reaction was theological. We were sinners in the temple of nature, at the altar of the mountains. Ah, but with our big home, fast food, and family road trips, the sin felt so good. I was torn between the guilt and the seduction of consumption.

As I got older, I felt more strongly that our civilization's growth mania was unsustainable and would likely end soon. Any day, the sky would fall. But somehow it didn't. Humans just kept driving more cars and building new homes. After high school, I played two years of ice hockey for the Billings Bulls in Montana before attending St. John's University in Minnesota. St. John's is surrounded by sugar maples and wetlands, and it is home to Lake Sagatagan (the inspiration for the fictional Lake Wobegon). I studied ecology there, and I fell in love with Aldo Leopold's poetic land ethic in his beautiful book *A Sand County Almanac*, which was written just across the border in Wisconsin. It was at college where I met a brown-eyed, guitar-playing girl named Amber. We hooked up on homecoming night in 1997, my freshman year, and we got married after graduation under the swirling limbs of a giant white pine tree on the edge of campus. Amber went barefoot, while I sweated with my groomsmen in tuxedos in the muggy Minnesota August.

After college, Amber and I volunteered for a year at an organic farm run by the Sisters of the Humility of Mary in rural western Pennsylvania. We actually lived across the border in Youngstown, Ohio, in a rent-controlled apartment complex the sisters ran in hopes of making a dent in the crushing poverty of another ghost town on the rust belt. There was a stray dog that lived in our parking lot. I named him Pile because he was a near-motionless heap of fur and bones. That dilapidated creature did a lot to symbolize Youngstown in those days.

Our second day on the farm was September 11, 2001. We were out harvesting potatoes when our new boss, Frank Romeo, a gentle old farmer whose wrinkled face seemed to be made from the very loam of the land, drove out to tell us a plane had just flown into the World Trade Center. Flight 93, which crashed near Shanksville, had flown over our backs as we stooped in the dirt on the side of the farm shaded by the mitten-shaped leaves of sassafras trees.

The following year we moved to Boulder, Colorado, so that Amber could train as a massage therapist and I could pursue a PhD in environmental studies. I wanted to restore the human relationship with nature, but I didn't know how. Environmental issues like sprawl, pollution, or climate change were too big for any one discipline. I thought of such problems as philosophical, but I knew that most philosophers weren't interested in actually getting their hands dirty with real-world issues. So I became a generalist and wandered around the academy, guided by my mentor, the wayward philosopher Bob Frodeman. I tried to glean useful bits of knowledge from wherever I could find them.

Despite this unorthodox training, in July 2006 I was offered a three-year postdoctoral fellowship in the Philosophy Department at the University of Twente in the Netherlands. I was in the parking lot of a Dairy Queen in Bessemer, Michigan (coming home from yet another one of Amber's cousin's weddings), when I got the phone call telling me the job was mine. We bought a Dutch-language instruction CD, and the whole ride back to Colorado we practiced saying things like *Hoe gaat het?* and *gezellig*. Over the next six weeks I scrambled to finish my dissertation.[3] Then we got on a plane.

In the Netherlands, I biked to campus every morning across a canal and through the Dutch countryside. Amber found us a three-story apartment above a jewelry store in downtown Hengelo, near the German border. Our old landlord, Meneer Jongepoerink, could remember being a young boy cowering under piles of rubble when the entire town was bombed to smithereens during World War II. In broken English, he recalled the story of seeing a bomb throw shrapnel over the head of one of his childhood friends, who miraculously survived unscathed. He was thrilled to rent the place to Americans, because it was American soldiers who passed out candy the day they liberated Hengelo from Nazi occupation.

The only building to survive the bombing was the old Catholic church. You can see pictures of the town circa 1944 where it is just the church surrounded by a flat sea of gray debris. A bomb actually crashed through the roof and slammed into the altar but didn't explode. Our apartment was a stone's throw from the church, along a street that turned into a bustling market every Wednesday and Saturday as tall Dutch men and stout Dutch women sold tulips, cheese, candies, and herring. On the morning of Ash Wednesday 2008, just as the vendors were setting up their stalls on the street below, our son, MG, was born in our home under the watchful eyes of two midwives. The first sound he heard was the clanging of the church bells calling the faithful to remember that they are dust and to dust they shall return.

MG learned to walk in the cobbled alleyway behind the jewelry store, and he learned to talk as he rode in a little seat mounted to the front of Amber's bike. Indeed, one of his first words was *fiets*, Dutch for bicycle. When he was eighteen months old, we moved to Denton, Texas, where I had landed an assistant professor position in the Department of Philosophy and Religion at the University of North Texas. It was during the recession, which made finding a job tough, but (as I would later learn) Texas had weathered the downturn better than other places thanks in part to the economic activity caused by fracking.

Going from Holland to Texas proved to be a bigger climate *and* culture shock than Amber or I expected. It was especially hard on Amber, who had fallen in love with the small size and slow pace of Hengelo—it reminded her of her high school days in the sleepy Northwoods town of Rhinelander, Wisconsin. Our first week in Denton, Amber, who had also grown fond of the quaint market on our street in Hengelo, had a nervous breakdown in the aisles of the enormous Kroger grocery store.

It was the fall of 2009. As we were looking for a home to buy, we'd often play at McKenna Park, which MG knew better as the purple park, because that was the color of all the playground equipment. That's where we first heard the word "fracking," when a young mom informed us that our new town was home to over two hundred gas wells. When she said this, it brought back memories of our family vacations to see Dr. Don's wells outside those shabby homes in Krotz Springs. Suddenly, the production side of fossil fuels was back in my life. Only this time it wasn't a vacation. This time the wells were in *my* back yard.

IN MAY 1841, about thirty years before Dr. Don was born, John B. Denton was killed by the Keechi Indians near a creek just east of the spot that would soon be known as Fort Worth. A pioneer lawyer, Methodist circuit rider, and Texas militia captain, Denton had been leading a posse to avenge a white family murdered by Indians. He was hastily buried in a shallow grave by the creek as his men retreated from the overwhelming Keechi forces. He left a widow and six children and became a local legend. In 1846 the new state of Texas named a county in his honor, and in 1857 its residents chose Denton as the name for the new county seat.

You can learn a lot about Denton from Shelly Tucker's haunted history and ghost tours every Friday and Saturday night on the downtown square. The way she tells it, in the early days the town of Denton was straight out of an old western film. Folks with true grit drank and gambled in wood-framed buildings that not infrequently burned to the ground. One of the first buildings on the square surrounding the courthouse, which sits in the center of town, was a hotel on the southeast corner lot. It had actually been operational in Alton, the old county seat seven miles south along Hickory Creek. The owner, who wanted to ensure he had the first hotel in town, put it on logs and used donkeys to pull it all the way to the square. The building inched along as each log it rolled over was taken from the back of the hotel and moved to the front. People came just to gawk at the absurdity of moving a building. The owner's pregnant wife reportedly sat on the front porch the entire trip, knitting baby clothes.

Denton's economy began with subsistence farming and ranching. In the 1880s, the railroads linked this small North Texas town to the national economy, and farmers switched to moneymaking crops such as cotton and wheat. Denton was even selected as a site for a national field laboratory in agricultural science. Experiments at that site helped to double the yield of several varieties of oats.

At the turn of the century, Denton became a regional center for higher education with the 1890 founding of the North Texas Normal College (now UNT) and the 1901 founding of the Girls College of Industrial Arts (now Texas Woman's University). Throughout this time, the courthouse, which sits on the highest point in town, underwent something of a Three Little Pigs evolution: the original wooden building burned down, its brick replacement was damaged by lightning and torn down, but the stately limestone courthouse that took its place still stands.

In 1901, John B. Denton's bones (which were identified by his gold teeth and a broken arm from when he fell off a horse) were reburied under the lawn surrounding the new stone courthouse. Unlike his first hasty burial, this final funeral was well attended. Throngs of men in suits and women in black mourning dresses crowded every inch of the square to pay their respects.

To this day, Denton's grave sits right there in the shade of the pecan trees that ring the courthouse, just across the street from Beth Marie's ice cream parlor, Jupiter House coffee shop, and the funky, labyrinthian used bookstore that occupies the old opera house. On her haunted tours, Shelly will tell you that some folks still swear that in the wee hours of the night they've seen Denton's ghost. He is said to wear a long black trench coat and carry a shotgun, pacing back and forth in eternal defense of his town.

As a southern city, Denton is also marked by the scars of slavery. A monument to Confederate soldiers on the south side of the courthouse square testifies to the ambiguous mixture of Texan pride and shame. Although Denton weathered the trials of racial integration in the 1960s and 1970s with unprecedented civility, it's not without its share of disgraceful stories about race relations. In the 1920s, for example, the city voted for a bond that was ostensibly to create a public park but was driven by the desires of white voters to remove an entire black community from the heart of the city near the Girls College. The bond effectively forced all the black residents of the Quakertown settlement to move their homes to a patch of land farther south infested with flies and mosquitoes and near an open sewer.

For the first half of the twentieth century, Denton hummed along as a small town just outside the gravitational pull of Dallas and Fort Worth to the south. Then came Interstate 35, the postwar boom in consumption, and a new ideal of urban development that was car-centric. In the 1960s, suburban sprawl—with strip malls, parking lagoons, and wide streets—started spackling itself around the old core of walkable neighborhoods filled with homes with front porches.

The house Amber and I bought was built in 1965. It straddles this divide between the old and the new Denton. It's still close enough to bike to UNT and the downtown square (and I do bike every day, a habit I acquired in the Netherlands and have stubbornly transplanted into the less hospitable environs of truck-loving Texas). But our home is also right next to a Whataburger and a dollar store on the six-lane Highway 380, which MG simply calls "the busy road."

Around the time our house was built, air-conditioning for homes and cars had gone from a luxury for the few to a standard expectation of the masses. In short order, having electrically manufactured indoor weather became a necessity, and Texans' in-home energy use spiked as they exchanged comfort for some of that true grit of the pioneer days. Air-conditioning also helped open up Texas to Yankee migrations. By 1970, Denton was the fastest-growing county in the United States. Today, the Dallas–Fort Worth area, with over 6 million people, is one of the fastest-growing metropolitan areas in the nation.

As a result, what was just a little farming community a hundred years ago now suffers from traffic jams and the worst air quality in the state. By 2030 the city of Denton is expected to nearly double its present population of 120,000. Yet despite this, Denton retains a quirky, small-town charm and a unique identity ("Keep Denton beard," as they say) sustained by a thriving music scene and creative class. It's a vibrant blend of little Etsy shops and Baptist churches that are so big the police have to direct traffic when services let out on Sundays. Many people affectionately call Denton "little d" to signify not just its smaller size relative to Dallas but more importantly its own sense of identity and defiant refusal to be swallowed by the metroplex. Residents drink wine and Shiner Bock and listen to live music on the square every Thursday night during the summer, laying out picnic blankets next to John B. Denton's grave and letting their kids run amok with their friends.

AROUND THE TURN of the twenty-first century, fracking came to Denton. The website for the Energy Information Administration has a time-lapse animation of drilling activity on the Barnett Shale, the massive reserve of natural gas that sits over a mile below Denton and much of North Texas.[4] The animation starts in 1997 with a small patch of black dots to the west of the Dallas–Fort Worth metroplex. The black dots represent conventional vertical oil and gas wells, the kind that had been used since the 1901 Spindletop gusher that christened Texas as oil country. For the first six years, the black dots slowly multiply, indicating a moderate growth in gas well development.

Then, around 2003, red dots rapidly proliferate on the animation. They represent the horizontal wells combined with hydraulic fracturing, the newly converged technologies that were about to turn the country into what the industry is fond of calling "Saudi America." Suddenly there is a bloom of red. As the animation runs from 2005 to 2010, it looks like an outbreak of measles rushing eastward and crashing like an avalanche onto the metroplex, which in turn is sprawling westward right into the heart of the gas patch. By 2014, there would be 17,000 gas wells on the Barnett, with another 14,000 forecast by 2030.[5] The production side of fossil energy, the very thing that made modern (air-conditioned!) development possible, had literally come home to roost.

The first place in Denton to experience the clash between fracking and development was Robson Ranch, a 2,700-acre gated suburb, which bills itself as an "active adult luxury retirement community." Robson is several miles southwest of downtown, separated by a pair of yet-to-be-developed ranches. It almost comprises a satellite town

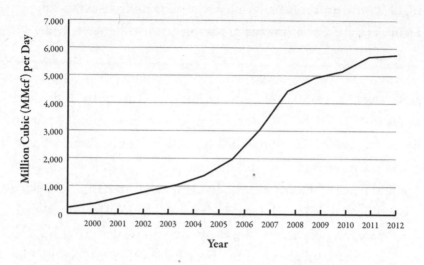

Natural Gas Production on the Barnett Shale, 2000–2012

With the perfection of hydraulic fracturing and its convergence with horizontal drilling and improved seismic imaging techniques, the Barnett Shale rapidly became the nation's most productive natural gas field. *Data courtesy Texas Railroad Commission.*

that exercises an outsized influence on local politics simply because almost everyone there is retired, reads the *Denton Record-Chronicle*, and votes. Its endless rows of nearly identical homes are straight out of the folk song "Little Boxes." Instead of ticky tacky, though, they are mostly made of stucco, a signature building material for the Arizona-based owner and developer Ed Robson.

By 1999, Robson had put together a master plan for his property that included gas wells alongside the homes, golf course, and elegant Wildhorse Grill. Some people would even seek out homes close to the gas wells, because Robson used the euphemism of "gas park" in its promotional materials, which sounded like it might be a nice place to toss the football around with the grandkids. Seeing these plans got city officials thinking about how drilling and development could coexist. With the Barnett Boom in its infancy, those blueprints gave the city a look into its future. Indeed, it was a look into the future of the country, as more than 15 million Americans now live within one mile of an oil or gas well.[6] In an attempt to prevent conflicts between neighborhoods and gas wells, the city created a gas well drilling and production ordinance in December 2001. At almost the same time, the city of Fort Worth implemented its own ordinance. They were the first of their kind on the Barnett, and arguably the first municipal ordinances designed to handle modern fracking anywhere on the planet.

Across the next eight years, companies like Devon Energy and Range Resources fracked nearly two hundred wells in the city limits with little controversy. Denton had annexed hundreds of acres, so most of the fracking was happening in western parts of the town that were still rural and sparsely populated at the time. Some conflicts here and there cropped up, which were handled the way most policy is handled, namely, by muddling through and satisficing. Even though the city was being pockmarked with industrial sites and approving dozens of gas well plats, each several hundred acres in size, there was no master plan, comprehensive studies, or guiding vision. It was all a series of one-off choices driven primarily by prices and private property with little thought about the bigger picture.

Then, in March of 2009, just as Amber and I were starting to plan our move to Texas, Range Resources obtained permits from the Texas Railroad Commission (the main regulatory agency at the state level) to drill five gas wells on the old Rayzor Ranch much closer to the inner core of Denton. The site they negotiated with the developer of Rayzor Ranch was directly across the street from a neighborhood, the Presbyterian Hospital, and the purple park.

The neighborhood mobilized in opposition, and dozens of doctors from the hospital signed a petition stating their objection to the project. Yet City Council members said they had no other legal option than to approve Range's request to develop three of the gas wells. Chris Watts, who was on City Council at the time, later told me it was the hardest decision he ever made as a public servant. "I felt like they had

a gun to my head," he said of the industry and the state laws that systematically favor its interests. If City Council didn't approve Range's
application, he and the other council members feared a crippling lawsuit would result.[7] Reflecting on all the problems that followed from
the Rayzor wells, Chris would later say that that decision was the one
he regretted the most during his six-year term on the council.[8]

The wells by the purple park were fracked in February 2010, the
same month MG turned two. Later that year, Josh Fox's influential
documentary *Gasland* was released, and suddenly it became apparent
that this seemingly local land use conflict was wrapped up with the
geopolitics of energy.

In light of the uproar over the Range wells on Rayzor Ranch, City
Council quickly moved to create a Gas Well Inspections Division
(before that, the fire department had been in charge of permitting gas
wells) and to strengthen the city's drilling and production ordinance.
Most importantly, they increased the setback distance between new
gas wells and existing homes from 500 to 1,000 feet (though operators
could still get variances to reduce that distance). But they also realized
the ordinance needed a more comprehensive overhaul, so they initi-
ated what would become a long and convoluted revision process. To
help them understand the technical issues and represent the "interests
of all stakeholders," the city created a task force. Three of its five voting
members, however, came from the oil and gas industry, which greatly
undermined its public legitimacy.

When the task force took shape in the spring of 2011, City Council member Kevin Roden approached me with his idea of creating an
unofficial "shadow" advisory body to counterbalance the industry tilt
of the official task force. Dallas was also in the process of rewriting
their ordinance, and a similar group had taken shape down there.
Kevin actually pitched his idea to a group of us philosophers who
worked at UNT's Center for the Study of Interdisciplinarity, headed
by Bob Frodeman and Britt Holbrook. I had been following the issue

ever since that day at the purple park when I first heard about fracking. This was my shot to get involved in a meaningful way as a philosopher in the field, so I agreed to help, and I formed the Denton Stakeholder Drilling Advisory Group.

I already knew Kevin a little bit by that point from chance encounters at the Civic Center pool, where our kids liked to swim together in the summer. He represented the relatively poor southeast district, where the Quakertown community had been resettled nearly a hundred years ago, and he was the favorite human jungle gym for all the children from that area who came to the pool to cool off. Whenever we got a chance to break away from the kids, we'd sit on the side of the pool and talk about his studies on his long road to a PhD in philosophy.

Politically, Kevin was fast becoming the darling of Denton—young, charismatic, intelligent, and devoted to his adopted hometown. Some of his concerns about fracking stemmed from stories beginning to emerge from his childhood home in North Canton, Ohio, about earthquakes near injection wells used to dispose of fracking waste. Kevin had gotten himself elected to City Council shortly after the wells near Rayzor Ranch were developed, thanks in part to his hardline stance on fracking. In one campaign event he promised to "nickel and dime the gas well operators to death with fees." The crowd erupted in cheers.

In June 2011, a month after Kevin was elected, Ed Robson approached City Council for approval of a plan to add more homes and a new gas well site to his development. The proposed gas well site would be right behind the home of a woman named Mitzie Fiedler, so she worked with her friend and fellow Robson Ranch resident Kathleen Wazny to oppose the plans. A couple of years earlier, Kathleen had successfully stopped Ross Perot's Hillwood Oil & Gas Company from developing a gas well site right next to her home. Shrewd and fierce, Kathleen actually shamed Perot into backing down by pointing out all the combat veterans in Robson who would now have to suffer the nuisances of an industrial site in their neighborhood. "Don't these men deserve a peaceful retirement?" she asked on a Dallas–Fort Worth news broadcast. A representative from Hillwood called the next day to tell Kathleen she had won.

But she lost her fight against Robson and his planned gas well behind Mitzie's house in 2011.[9] This time, the community turned against Kathleen. As she would later tell me, "Many of the folks in Robson Ranch are good old Texas boys who love the oil and gas industry." They flooded City Council with e-mails in support of Robson's plan, and they aimed their hostilities at Kathleen. "Things got really ugly," she told me. "I was afraid for my safety. . . . I was sure someone was going to vandalize my house." Ed Robson also had a notice sent to all the residents of his development that if his new plan wasn't approved, their homeowners association fees would have to be increased. The note explained that gas well royalties subsidized the pool, community center, golf course, and fitness center. The thought of losing those amenities left a lasting impression on Robson's residents. In the end, the gas well site was developed, Kathleen licked her wounds, and Mitzie packed up and moved to Florida.

Two months later, on the night of August 25, 2011, the official task force held its first meeting.[10] City staffers had set out about a hundred folding chairs for the audience in the cavernous round space of the Civic Center. It was our ten-year wedding anniversary, but there Amber and I were, sitting in the front row, trying to get comfortable in those metal chairs. And there was MG, who was three years old by then, sitting cross-legged playing a video game on Amber's phone. None of us knew it at the time, but Amber was newly pregnant. So there too was little Lulu, somewhere small and deeply hidden away. Louisa Mae, the only native Texan in our family, would be born in May 2012, just as the Denton mayoral election went into a runoff, driven in large part by the question of urban fracking.

————

THREE HUNDRED THIRTY million years ago, North Texas was submerged under a tropical sea that slashed its way up through the continent into present-day Pennsylvania. For 25 million years, tiny protozoa swam among blooms of algae and darting squids. Mollusks clung to rocks on the water's edge, and starfish inched along the bottom of the sea. The land was so teeming with ferns and ancestral trees that oxygen levels in the atmosphere spiked, allowing for the growth of giant insects. Dragonflies had two-foot wingspans. Spiders were the size of cantaloupes.

But the bottom waters were oxygen starved, meaning that all the bits of ferns washed into the sea by thunderstorms and all the carcasses of radiolarians, algae, and cephalopods that rained down to the seabed were preserved. Then eons passed—enough time for entire mountain ranges to rise and fall and rise again like the rocky lungs of a breathing planet. And all this time the bodies of those sea creatures were slowly pressure-cooked into mudstones, their carbohydrate innards converted into methane. By the time John B. Denton was shot and buried in a shallow grave, this ancient sarcophagus had become a solid black rock laced with invisible pockets of natural gas.[11]

The Barnett Shale is at its thickest and richest just under Fort Worth and the western half of Denton, but then it abruptly ends along a fault and a massive underground geological formation known as the Muenster Arch, which aborted the shale's production of methane when it came crashing upward. The subterranean Muenster Arch runs somewhere roughly under Bonnie Brae Street in Denton. "Bonnie Brae" is Scottish for "beautiful hill," a name given to it by Charlie Wilkins, one of the earliest settlers in Denton and a member of the Confederate army.

There is indeed a beautiful hill along Bonnie Brae; I jog there often in the mornings when the post oaks still cast long shadows across the sleepy avenues. On the western slope of that hill is the purple park. If you sit in the swing there facing west toward the newer, rapidly growing parts of Denton, you'll look out on land pockmarked with some of the biggest producing wells on the shale. But if you turn around and face east down Scripture Street, toward the courthouse square and the older half of town, you'll be looking at land with no shale underneath it at all. Our house by the Whataburger about a mile east of Bonnie Brae is probably just on the other side of the shale. The closest well to us, the Parks Unit #1H, stretches its horizontal pipe underground just to the edge of our neighborhood. There are no active Barnett Shale gas wells to the east of our home.

If you looked south of Denton and could see underground, you'd watch the Barnett Shale slowly thin out and rise like a submarine to the surface until it emerges along a stream in San Saba County two hundred miles away. In the latter part of the nineteenth century, that stream flowed through land the government had given to Captain John W. Barnett, a contemporary of John B. Denton, as a reward for his service in the Mexican War for Texan independence. When he settled there, Barnett named the stream after himself.

Of course, Barnett and Denton knew nothing about the geology of shale, and they would have had little use for it even if they could have unlocked the gas inside. They did not lead lives, like ours, that are dependent on industrial chemicals and fossil-fueled energy slaves. But that changed quickly. In December of 1912, Denton residents started receiving natural gas for the first time. A sixteen-inch pipeline owned by Lone Star Gas Company brought gas from the Petrolia Field ninety miles northwest of town and fed it into the homes newly equipped with gas-burning stoves.[12]

As demand grew, geologists began scouring the country to catalog the oil- and gas-bearing rocks. When they discovered the shale on a stream in San Saba County in the early twentieth century, they naturally named it after Captain Barnett. They knew the rock contained methane, which in theory could be harvested. But for decades, shale formations like the Marcellus and Barnett were largely forgotten, because everyone assumed there could be no economical way to recover the gas. It was locked up way too tight in a rock that is blacker and harder than a hockey puck.

Then came the energy crisis of the 1970s and President Carter's speech about how the country was running out of oil and gas and how we'd been living high on the hog, but now the party was about to end.

Except that it wasn't about to end. For even as Carter was making his doomsday speech in Washington, DC, teams of geologists and engineers were tromping through the nearby woods of West Virginia in search of a technological fix. The Energy Research and Development Administration (precursor to the Department of Energy) and other federal agencies had partnered with several private companies to run a massive R&D effort known as the US Unconventional Gas Research Programs.[13]

The scientists and engineers behind this program knew that the United States wasn't running out of oil and gas; it was just running out of the stuff that was easy to reach. The traditional, easy stuff would collect in underground reservoirs that were relatively loosely bound, such that if you stuck a wellbore in them you could pull out the oil and gas just like sucking soda up with a straw—indeed, many wells flowed upward under their own pressure. Those reservoirs originated from "source rocks" like the Barnett or the Marcellus. Such tight formations are often called "the kitchen," because that's where the oil and gas are originally "cooked."

To achieve a new dawn in American oil and gas, engineers would have to figure out how to extract resources directly from the kitchen. That's why the shale formations that were nearly forgotten became a top priority as energy research on fossil fuels grew by an order of magnitude between 1974 and 1979. The goal was simple: squeeze, or rather frack, more oil and gas out of American rocks. There was no time to wait for oil and gas to migrate out of the kitchen into more easily tapped reservoirs. We needed the energy fix now. You could call it the beginning of the age of extreme energy. It's when, just as we were starting to understand anthropogenic climate change, we doubled down on fossil fuels and went to ever greater lengths to extract and combust them.

Then again, we've been going to extremes for a long time. If we define fracking broadly as breaking up oil- and gas-bearing rocks, then it has been around since the 1860s, when drillers in Pennsylvania started experimenting with explosive nitroglycerin to increase oil production. It often worked, but it made an already dangerous job even more perilous. The greatest experiment in explosion-based fracking came a century later when, in 1967, the US Atomic Energy Commission detonated a 29-kiloton nuclear bomb 4,200 feet below the surface of northern New Mexico in an attempt to liberate shale gas.[14] This "Project Gasbuggy" did increase production from the Lewis Shale formation, but (surprisingly?) the gas was too radioactive to be useful.

The genesis of modern fracking was in the 1930s, when engineers started using nonexplosive fracking techniques that called for pumping fluids down the well at high pressures. They tried lots of recipes, with varying degrees of success—acids, thickened gasoline, diesel, gels, water, and foams. In 1947, Stanolind Oil and Gas Company conducted the first experimental treatment of a nonacid hydraulic fracturing (called Hydrafrac) in Kansas, using 1,000 gallons of a thickened gasoline. That experiment was largely a failure. But engineers kept tinkering, and by 1950, two years before the Don Deal well came into production, Halliburton Oil Well Cementing Company had hydraulically fractured 332 wells, with an average production increase of 75 percent over conventional (nonfracked) wells.[15]

But these techniques were not yet powerful enough to raid the kitchen rocks where the real treasures lay waiting. Throughout the 1980s, the experiments with horizontal drilling in West Virginia kept running into barriers. Think of the shale as the creamy filling of an Oreo cookie. Traditional vertical wells had just stuck a straight straw into that filling. They hit the pay zone but couldn't suck up much of the good stuff. By drilling down and then turning a bendy straw horizontal for a while (now up to a mile or more), engineers can vastly increase surface area contact with the shale. This is not as easy as the bendy straw analogy makes it sound, though. You have to pilot a drill pipe through the pitch-black heart of the rocky earth and get it to land in a very narrow pay zone at depths and temperatures that turn the pipe into a flailing string of spaghetti.[16]

Other obstacles were thwarting efforts in the 1980s and 1990s to perfect a more powerful form of fracturing. It was difficult to get enough pressure and to find the right mixture of sand or other proppants to hold open the induced cracks in the rocks so that the oil and gas can flow. It wasn't clear which of a variety of gels or foams (or even water) would be the best medium. And there was constant experimenting with a vast array of chemicals to do a variety of things from killing bacteria to lubricating flow. Solving these problems would take the dogged efforts of the son of poor Greek immigrants working on wells just outside of Denton.

George P. Mitchell was born in Texas in 1919, the same year my great-grandpa bought his swampland. Mitchell graduated from high school at the age of sixteen and worked his way through Texas A&M University selling gold-embossed stationery to fellow students so that they could write love letters to their sweethearts back home. He was captain of the tennis team and finished first in his class in petroleum engineering. He went off to work for Amoco in the oil fields of Louisiana, though by the time the Don Deal well started producing, Mitchell was back in Houston working with his brother Johnny, usually at a drugstore counter, to start up an oil company of their own.[17]

One of the first things the Mitchell brothers did was lease over 300,000 acres in north Texas.[18] Back then that area, which is identified geologically as the Fort Worth Basin, was known as "the wild-catters' graveyard," because everyone was striking out with dry holes. But Mitchell seemed to have what scientists call "golden hands," that knack for getting an experiment to work when no one else could. By 1964, Mitchell and his company had resurrected that graveyard with a thousand successful wells. The good years followed, supplying gas to a growing Chicago market. But by the late 1970s, it was clear that their main gas play, a conventional reservoir known as the Boonsville Bend Conglomerate, was being inexorably depleted.[19]

Yet whereas President Carter advised putting on sweaters in the winter, Mitchell started looking for ways to keep the gas flowing. When his engineers told him in 1980 that they might only be able to milk another ten years' worth of production out of the Boonsville, his response was calm and clear-eyed: "That gives us plenty of time to find a replacement."[20] For as large as his company had grown, it was still no Shell or ExxonMobil, so he couldn't just pick up shop and move overseas. All Mitchell could do was dig about 2,000 feet under the Boonsville into the Barnett and pray he could make it work.

In 1981, about thirty minutes southwest of Denton, Mitchell drilled the C. W. Slay #1 gas well into the Barnett, and the following year he connected it to a pipeline. That makes it the discovery well for the Barnett.[21] Given that the Barnett was the first shale play to be developed, you might consider the C. W. Slay to be the omphalos of fracking, civilization's first umbilical cord into a new supply of nutrients.

Yet it took another sixteen years and $250 million in investments to really unlock the Barnett. The C. W. Slay and other early wells produced only a trickle, partly because the operators weren't able to do horizontal drilling and partly because they couldn't figure out a profitable way to fracture the shale. For years, Mitchell dragged his company along behind him to an increasing chorus of complaints that he was wasting their money on this fool's errand of shale gas. He tried experiment after experiment looking for the right recipe for fracturing. At one point, a skeptic told him, "If the Barnett is the best thing we have, then we don't have shit."[22]

But in 1998, just when the money was running out at Mitchell Energy, a young engineer stumbled on the right formula at the S. H. Griffin #4 well. At the same time just a few miles to the east, developer Ed Robson was putting the finishing touches on his master plans, complete with golf course and "gas parks," for his active retirement community.

Today, the S. H. Griffin is just one of hundreds of wells interspersed among small prefab homes with large lots for horses. It's near a playground and the aluminum siding building that serves as the town hall for Dish, the little community formerly known as Clark that sold its name in exchange for free satellite television. Around its perimeter is a flimsy chain-link fence. Inside that, you'll find the usual parts: the condensate tank (which temporarily stores produced liquids that come up with the methane), separators (tubes that separate the liquids and gases), and the "Christmas tree" (the assembly of valves and fittings used to control the flow of gas). It's all contained in a knee-high swath of corrugated metal, giving the impression that the gas well is wading in a homemade kiddie pool. It's bizarrely prosaic considering that this gas well is arguably the epicenter of a global energy revolution.[23]

For a long time, Mitchell's team had assumed that breaking open the Barnett required gels or foams. But as the S. H. Griffin showed, all they needed was water, millions of gallons of it, mixed with the right concoction of chemicals and sand and forced under high enough pressures. The Barnett suddenly boomed. From 1998 to 2007, gas production increased more than thirty-fold from 94 million cubic feet per day to over 3 billion cubic feet per day.[24] They had finally "cracked the shale code," and Mitchell—who grew up in Galveston in a modest space above his father's cleaning business—had become the "father of fracking" and would be compared to Edison, Ford, and Tesla in terms of his role in making world-changing innovations.[25]

In the three years after the success at S. H. Griffin, Mitchell Energy grew by over 250 percent. In 2002, the year after Denton wrote its first gas well ordinance, Devon Energy bought the company for $3.1 billion and married Mitchell's recipe for hydraulic fracturing with the horizontal drilling techniques that were then starting to be perfected by engineers working back east. A few years later the first rumblings of an energy boom would be heard along the windy plains of North Dakota and in the wooded hills of Pennsylvania where Amber and I once harvested potatoes. In 2013, Boulder City Hall would be overrun by antifracking protestors, even though no one ever uttered the word "fracking" when Amber and I were at school there just a few years earlier. And by 2014, thanks to fracking on the Eagle Ford Shale not far from Mitchell's hometown, Texas would surpass Iraq to become the world's sixth-largest producer of oil.[26] Indeed, the entire United States would witness a rebirth in domestic crude oil production.

Largely as a result of what transpired just outside Denton, the United States has now gone from President Carter's gloomy forecast of scarcity to having so much natural gas that prices have plummeted and facilities designed to import it are being retooled to export it. Globally, innovation has unlocked shale deposits that hold over five times the amount of oil the human species has consumed throughout history, fending off dour forecasts of "peak oil" (the point at which oil production enters a terminal decline).[27] This boom has brought with it jobs and economic growth; by 2014, fracking had sparked a resurgence of manufacturing and drastically cut the unemployment rate in Youngstown, revitalizing a place that Amber and I knew as a ghost town.[28] For America, the boom brought the promise of energy independence, the holy grail of national policy since the oil shocks and gas lines during Carter's presidency. In China, fracking brought the hope of cleaner air from less coal combustion.[29]

But fracking has also brought concerns. In 2011, when Denton began rewriting its drilling ordinance, Calvin Tillman, the mayor of

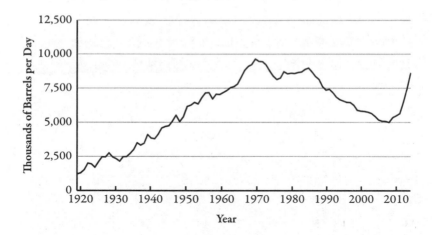

US Field Production of Crude Oil, 1920–2010

Note how around 2010 the fracking revolution reversed a steady decline in US oil production since 1970. The production totals for October 2014 were the highest recorded since May 1985. *Data courtesy US Energy Information Administration.*

Dish, was forced to leave his own town. His sons' asthma and nose-bleeds kept worsening as the gas wells and pipelines kept prolifer-ating. Around the same time, state authorities in Pennsylvania had ordered Cabot Oil and Gas to supply potable water to ten families in Dimock whose water wells had been contaminated shortly after nearby fracking began. Two years later, as Denton's City Council voted to adopt their new drilling and production ordinance amid great protest, teachers in the rural schools of South Texas a few hun-dred miles away would begin carrying hydrogen sulfide monitors in case toxic gas levels should spike from the surrounding array of oil well flares.[30] And in the first half of 2014, Oklahoma, which had been riddled with injection wells to dispose of fracking waste, surpassed California with 207 earthquakes above magnitude 3.0. According to the historic average, that's a hundred years' worth of quakes in just six months.[31]

CHAPTER 2

CLOSING THE BARN DOORS

THESE THREE STRANDS—MY OWN BACKGROUND, the City of Denton, and the development of fracking—all braided together in the story of the Denton Stakeholder Drilling Advisory Group (or DAG), the collection of citizen advisors that I put together at the request of City Council member Kevin Roden. We worked in parallel with the gas well advisory task force, the group with a majority of industry members created by the city to help rewrite its drilling ordinance. The task force was an official city entity with all the trappings that entails—weekly meetings held on the dais at City Hall, access to city staff and resources, and even closed-door briefings with city attorneys. By contrast, DAG had no official standing as a board or committee. We were just a collection of citizens, but in the end we left a larger and more lasting imprint on the politics of fracking than the task force did.

Fracking, especially in urban areas, suffers from a split-personality disorder, being simultaneously about energy and land use. From the first perspective, natural gas development is compared with coal, wind, and solar power. But from the second perspective, the relevant comparisons are to payday lenders, adult video stores, and paint shops. Fracking represents both our global interdependence on networks of commodities and our place-bound lives in communities.[1]

The first perspective is the stuff of state and national politics. The second perspective is the local one. Talk to any Denton City Council member or city staffer about fracking, and it won't be a discussion about climate change and energy independence. It'll be a conversation about zoning, flood easements, neighborhood integrity, plats, and permits.

The question confronting Denton was how to find the optimal balance between two sets of competing goods. On one hand there was the hope of fostering revenues for the city in the form of jobs, taxes, and royalties, as well as an obligation to protect mineral property rights (the rights of the people who own the "mineral estate," namely, the creamy filling of the Oreo where the gas is trapped a mile below the surface). On the other hand, there was the need to protect the health, safety, and welfare of Denton families who live above on the "surface estate" and are exposed to the risks and harms of the intensely industrial process of extracting the minerals from below.

I would eventually come to call this the compatibility strategy, as the overarching goal was to make fracking compatible with communities. Think about an elementary school in a residential neighborhood. This can be a compatible land use for that area, so it is technically permitted in Denton's city code. But there are good reasons for concerns (e.g., about increased traffic). So, city officials need to examine this on a project-by-project basis to work out any potential clashes. To do so, they employ a specific use permit (SUP) process that entails a public hearing and involves writing a special set of rules to make the school compatible with that neighborhood.

When it comes to gas wells, however, the situation is muddled. Gas well sites are also technically permitted in every zoning category, including residential areas. But this is not because (as with schools) anyone really thinks gas development is compatible with neighborhoods. Rather, it's because the mineral estate is superior to the surface estate. Given the strong legal standing of mineral rights, many cities have been concerned that not permitting gas well development everywhere would trigger a lawsuit. By allowing gas wells in every zoning category, Denton's city code largely answers the compatibility question by default: gas well development *is* compatible with all other land uses. Now, gas well operators still need to get an SUP to do business in Denton, but city officials could not deny a permit application on the basis of the claim that drilling and dwelling are simply incompatible. The default presumption of compatibility meant that the city was confined to relatively minor rules to mitigate some of the nuisance issues (like aesthetic improvements of requiring landscaping and sound barriers). This frustrated City Council, because the public hearing process for the SUP made it look like they had the power to deny a gas well, but they really had no such power.[2] They could only put lipstick on the pig, not keep the pig out of back yards.

As a result, even though gas wells are classified as industrial land uses in the city code, they can be located in any zoning category, including residential.[3] That's why fracking is a uniquely invasive industry. The city doesn't even allow bakeries in residential areas, because they are considered incompatible land uses (with truck deliveries and early morning activity). Nonetheless, as Denton rewrote its ordinance from 2011 to 2013, the question was not really on the table about *whether* to continue to allow fracking in neighborhoods. The only question was, given that it is going to happen, what set of rules (for the SUP) would make it less bad.

Even on the oil-and-gas-friendly task force, no one disputed that fracking is an industrial activity. On average, it takes 1,200 diesel trucks to bring a single well into production.[4] The drilling process can continue for a month or more, with loud noises and the constant revving of giant diesel engines belching plumes of black smoke.[5]

Hydraulic fracturing can last over a week, with yet more noise and light pollution and sand, diesel fumes, and other odors wafting off-site.[6] The flowback process, when many of the chemicals shoved down the hole come gushing back up in front of the methane, can take only a couple of days but is also often intensely noxious.[7]

Things quiet down significantly after flowback when the gas well is in production, but there is still the possibility of leaks and spills, as well as the regular truck trips to empty the condensate tanks.[8] There are also chronic emissions of toxic chemicals from those tanks. Even the fine print of industry documents notes that the business poses "risk of fire, explosions, blow-outs, pipe failure, abnormally pressured formations and environmental hazards such as oil spills, natural gas leaks, ruptures or discharges of toxic gases."[9]

There are conflicting scientific reports about the health and environmental impacts of all this industrial activity.[10] What risks does fracking pose to ground and surface waters? How much does it contribute to air pollution in the Dallas–Fort Worth area? These were live questions at the time DAG and the task force were working simultaneously to recommend a new gas well ordinance. And because fracking was already up and running, any changes to the rules required some evidence that problems existed. So as uncertainty and conflicting studies reigned, groups like DAG had an uphill climb in trying to prove the case for more robust safeguards and regulations.

But I thought that by 2014, the scales were beginning to tip as more and more scientific studies found problems caused by fracking. A report released by a group of physicians in July 2014, for example, drew from a database of hundreds of peer-reviewed studies to argue that there is "no evidence that fracking can be practiced in a manner that does not threaten human health." Citing "rapidly expanding evidence of harm" and remaining "fundamental data gaps," the report recommended a moratorium on fracking as "the only appropriate and ethical course of action."[11] Many of the studies this 2014 report cited were published during or after the work of DAG and the task force in 2012. That makes me wonder if things would have turned out differently had Denton's policy revision happened two years later.

ALTHOUGH NO ONE on the task force disputed the industrial nature of fracking, none of them had actually lived through it in the way Cathy McMullen had. A nurse, Cathy spends most of her life caring for patients suffering from cancer and other serious illnesses. In her spare time, she volunteers at the Denton animal shelter, where she rescues as many dogs as she can handle, especially ones in need of medical attention. I remember in particular one long stretch of anxiety for Cathy when her newly adopted Rottweiler kept ripping open stitches in his chest. Cathy agonized and doted over him, just as she does for her patients. Even during the intensity of the campaign for the fracking ban, Cathy found the time to care for a scrawny stray cat. In fact, she took such good care of Ninja Kitty that she soon had to put her on a diet.

Cathy's eyes carry that deep and soft sadness that comes from inviting so much into your heart and having to say good-bye so often. But so much loss has also given Cathy a hard edge, which she has come to rely on as Denton's most battle-hardened critic of fracking. In 2008, she and her husband, Ron Watson, a retired FBI agent, were living on eleven acres out in the country near Decatur, about thirty minutes west of Denton. It was an ideal place to raise and care for animals. But then Aruba Petroleum began drilling near her property. At one point, eighty-eight frack trucks were lined up outside her home, full of water, sand, and chemicals. Aruba tore down her plum- and peach-tree orchard for their pipeline. Her neighbor's cows died after drinking drilling fluids. During fracking, the air hung heavy with diesel odors and other acrid smells.

Cathy did not own the mineral rights under her land, so she had no say in the matter and she got nothing for her troubles. One day, channeling the surly persona she'd acquired from years of tending bar, Cathy marched out to the front of her property and refused to let the trucks come on her land. She brought her biggest dogs with her, and she made her stand until the sheriff came out and told her that she had to let them through. The law was on their side. Cathy felt she had no choice but to move. Her home had been devalued, but she was able to sell it to someone whose company was paying for him to stay there temporarily. Her neighbors Bob and Lisa Parr were not so lucky. Cathy remembers seeing Lisa a few years later and being floored by how sickly she had become, likely from the surrounding fracking.[12]

When Cathy found a nice house in Denton near the purple park with a beautiful front porch and a big yard for dogs, she made sure to ask the real estate agent if there were any gas wells in the area. She was told that she would not get the mineral rights with the house, but that there was no gas development nearby anyway. Shortly after she moved, however, Range Resources applied to Denton City Council for permits to frack the three wells on Rayzor Ranch near the purple park. Soon she saw the posts with little pink flags, the telltale sign that a fracking operation is fixing to move in. Cathy got mad. And she got busy—appealing to City Council to deny the local permits, organizing her new neighborhood, hiring local attorney Sara Bagheri, and raising funds to gather baseline air quality samples. She's the one who got the doctors from the nearby hospital to sign letters of opposition.

As she talked to her neighbors, Cathy heard of another woman who had recently moved to Denton after a bad experience with fracking west of town. Cathy began reading this woman's blog, which was just starting to make waves in the region. It didn't take long for her to contact Sharon Wilson, better known now as "Texas Sharon," one of the nation's most prominent fracking foes. Sharon advised Cathy and her neighbors about how to conduct air testing, record their health symptoms, and register complaints with state authorities. She also strategized with them and Sara about how to push for tougher regulations at the city level. Throughout the ordeal, Cathy and Sharon came to develop a close friendship, one that was strengthened by their shared history with fracking.

Sharon spent her childhood at her grandparents' farm just outside of Fort Worth, where she learned roping, riding, and shooting. She worked as an administrative assistant for a couple of oil and gas companies until 1996, when she bought forty-two acres in Wise County just west of Denton to raise her son Adam, ride horses, and live off the land. Soon she found herself at the epicenter of the shale boom. Sharon did own mineral rights on her land, so when the landmen came to her door with talk of big money, she signed the lease with great excitement about the prospects of a brighter future for Adam. She even wrote a letter to Aruba Petroleum, asking them to drill on her property and included a map marked with an X on a spot by the road that she thought would be a good place to put the frack site. But one day in 2005 her tap water turned black and slimy.[13] Adam went without his bath that night while Sharon tried to hide her fear from him. As the rural landscape started to industrialize with gas wells, she started educating herself about the downsides of fracking. She found that foul water was only a piece of a much larger array of problems. After countless hours of activism on her own time, the national non-profit environmental group Earthworks hired her in 2010 to be their leading voice on fracking.

Along with Josh Fox, Sharon has come to occupy the center of the sprawling web of antifracking activists. The *Fort Worth Weekly* called her "one of the industry's worst nightmares."[14] She can give you the sweetest smile under her short, sandy blond hair, but she can also squinch her eyes into a laserlike death stare. When I first saw that look I made a mental note to myself *never* to be on the receiving end of it. Sharon knows all the big names in the fracking scene but also every mom in North Texas fighting for her kids' health in city halls.

She runs the Bluedaze blog, which she has doggedly crafted into the leading site for whistle-blowing, muckraking, and general hell-raising about all things fracking, including her courtroom tussles with what she calls the "Big Oil and Gas Mafia." [15]

One cold, clear day in February 2010, Cathy was working at the Presbyterian Hospital and looked out the third-floor window to see a giant flare coming off the gas wells at Rayzor Ranch. She immediately called Sharon and Alisa Rich of the nonprofit consultancy Wolf Eagle Environmental, who took an air sample with a summa canister, a small metal vessel with a valve that can be opened to draw air inside and trap it so that it can be analyzed at a laboratory. The results would later show that benzene, a carcinogen, was at unsafe levels, something Cathy had suspected from the sickening odors coming off the site.[16] Cathy also walked over to the purple park, where she snapped a picture of the flare framed in the foreground by the playground equipment. For me, that became an iconic image of the clash between fracking and neighborhoods. How were we going to make them compatible?

It wasn't until about a year and a half later that I met Cathy and Sharon. Kevin and his wife, Emily, host "drink and think" events in their spacious Victorian home near the campus of Texas Woman's University. They are adaptations of Parisian salons, where intellectual exchange about contemporary issues is lubricated by Armadillo Ale and other local craft beers. It's one of those "Denton things" that makes it such a special place. One night in July 2011 Kevin modified a "drink and think" into a screening and discussion of *Gasland*. After the film, he invited me to come to the front. I told the crowd of fifty people that I was starting DAG in order to offer more citizen input into the rewrite of the drilling and production ordinance. Cathy and Sharon were the first to sign up. Several others joined too, including Elma Walker, a retiree at Robson Ranch who had been active with Mitzie and Kathleen in the fights over fracking down there. Elma had memorized all the ordinances from the various cities atop the Barnett

Shale and could cite, almost off the top of her head, the clauses she thought Denton should adopt. Elma and her husband, Bruce, were regulars down at City Hall, where she would flit around like a bird, politicking with the movers and shakers, while he would plod like a turtle telling long stories to anyone willing to listen.

The idea behind DAG was not just that the people *cared* about the issue, but that some of them *knew* more or different things than the credentialed experts who got to sit on official advisory bodies like the city's task force. Everyone in the group had what might be called in other contexts "indigenous knowledge." Texas Sharon is an example. The kind of knowledge she has did not come out of textbooks or degree programs. It came out of her kitchen sink in the form of polluted water. Add the Internet to this experiential knowledge and you get a citizen-expert. And, in Sharon's case at least, you get a real pain in the ass for the industry.

DAG was created to counterbalance the industry tilt of the task force. It wasn't hard to predict that the task force was going to come up with some fairly superficial changes to the ordinance. The members of DAG, by contrast, wanted to push for more robust protections for safety, health, and welfare. This presented the defining challenge for DAG. We had to be more radical than the official institutional channels like the task force while at the same time garnering enough legitimacy to get people in power to take our ideas seriously. Slide too far one way and you are consigned to the lunatic fringe (whether you deserve to be put there or not). Slide too far the other way and you become just another cog in the intrinsically moderate bureaucratic machine.

Our strategy was to host public forums featuring prominent voices and hope their standing would rub off on us, giving our group some heft in the public policy process. In August 2011, DAG held its first forum on the UNT campus to a standing-room-only crowd. The first speaker was the financial analyst Deborah Rogers, who had suffered health problems including nosebleeds when fracking happened near her Fort Worth home. Deborah's likening of shale gas to a giant Ponzi scheme had recently made the *New York Times*.[17]

The headliner was Calvin Tillman, former mayor of Dish, the town that hosts the S. H. Griffin well in the heart of the gas patch just west of Denton. Calvin gave a powerful speech and concluded with his own story of being forced to leave Dish because of concerns about his children's health. He said, "I used to be a Republican like these people [members of the oil and gas industry], but they have no respect for property rights, which is supposed to be the bedrock of conservatism. . . . I mean, you look around at what happened in Dish, to my neighbors and family, and you wouldn't believe this is America."

DAG held two more public events on the UNT campus in October 2011, which both drew good crowds despite competing with the Texas Rangers' epic and ultimately heartbreaking playoff run (though not so heartbreaking for my dad and other Cardinals fans). One featured scientists on the air and water impacts of shale gas development; the other, on the regulatory landscape, included the executive director of the Railroad Commission (the main oil and gas regulatory body in Texas) and a completion and construction manager from Devon Energy. MG, who was three years old at the time, would tag along with me to coffee shops and restaurants to help hang up flyers advertising DAG events.

DAG started getting newspaper headlines as our strategy worked. The task force even postponed their deliberations after their first meeting to allow more time for DAG's public education events. As I listened to this diverse range of speakers, one overriding message impressed itself on me: no one really knew much about the environmental and public health effects of fracking. The science was immature and inconclusive, leaving plenty of room for dueling experts and for interest groups to cherry-pick the findings that suited their positions.

The government R&D programs and Mitchell's heroic efforts had all been focused on maximizing the recovery of gas at a minimal cost. Very little money had been spent to foresee, track, and catalog the cumulative impacts of such a massively scaled-up extraction process. Engineers had built into the equation certain values like efficient gas recovery but had left out other values concerning water and air quality or neighborhood livability. I began to realize that the ordinance revision process was also a form of engineering. We were trying to build a wider set of values into the system. Mitchell did the innovation; we were doing the renovation.[18]

We could have spent months hosting talks, but eight new permits to frack in Denton had been issued during the time we were scheduling public forums. Public education was great, but it was butting heads with the need to improve the ordinance quickly. Feeling the pressure to act (the task force had announced they would resume their official deliberations soon), in November, DAG issued a one-page statement of six values comprising what we called the principle of responsible resource development. A month later, we issued a full report.[19] We made over forty recommendations for building more values into the ordinance, including ideas for reducing air emissions and noise, protecting ground and surface water, managing waste, stewarding gas well sites, and preventing accidents.

Overarching all of this, we argued that the city should enact a moratorium on new permits. Remember, that's the precautionary principle: no fracking until we prove it's safe. The language of precaution was initially the suggestion of DAG members Ed and Carol Soph, musicians who had moved to Denton in the late 1970s in hopes of finding steadier pay than the freelance scene in New York City could offer. When they got to Denton, Carol set down her French horn to raise their son, while Ed worked his way in at UNT's world-renowned jazz studies program, first as an adjunct and then as the nation's first tenured professor of drums. Ed and Carol lived in a neighborhood known as Idiot's Hill, pretty far east of the shale, but they had experience successfully fighting a copper smelter that wanted to set up shop, and emit lead, in Denton in the late 1990s. They saw the fracking issue as similar, yet more serious in terms of its environmental and public health implications, to the copper smelter fight.

Ed and Carol were firm believers in the precautionary principle, and we all agreed that we had learned enough to see that urban fracking was happening too fast with too much uncertainty about the potential harms. As the task force got set to resume their deliberations, the *Denton Record-Chronicle*, known locally as the *DRC*, put us on the front page with the headline "Caution Is Their Watchword." [20] My ninety-three-year-old neighbor, Miss Pearl, was so tickled to see my picture in the *DRC* that she walked across the street, something she hadn't done in years, to give me her copy of the paper.

AFTER WE ISSUED our report on the ordinance, the task force began weekly meetings at City Hall that lasted through the spring of 2012. They met on Monday nights, sitting up at the dais occupied by City Council members on Tuesday nights. The task force began by marching through the DAG report, discussing each of our recommendations point by point. Public attendance was high at first, but attrition slowly set in until by the end the audience was pretty much just me and Peggy Heinkel-Wolfe, the local *DRC* reporter. Listening to discussions about insurance requirements and landscaping codes was a lesson in how municipal bureaucracy can make even a controversial issue like fracking boring.

Vicki Oppenheim, an environmental planner and leader of Denton's community market, was joined by Tom La Point, a UNT biologist who studied water quality and aquatic ecosystems, as the nonindustry members of the task force. Sitting next to Vicki was Don Butler, a project manager for a consulting firm for the oil and gas business. To his left was the most notorious member, Ed Ireland, who was head of the

Barnett Shale Energy Education Council, an industry-funded public relations outfit. Dr. Ireland's bald head and dark, deeply set eyes gave him the intimidating visage of the archetypical cigar-chomping business tycoon. Though always cordial, he did not disguise his dislike for antifracking activists, and the feeling was mutual. When I invited him to debate Texas Sharon in my graduate seminar on energy policy, he called the invitation "an insult," because "you equate my expertise as a Ph.D. economist and a 25 year veteran of the oil and gas industry with Sharon Wilson, who is a secretary-turned-environmental activist." [21]

Sitting next to Dr. Ireland was John Siegmund, a retired oilman with a slow Texas drawl and a smile as wily as J. R. Ewing's. He had worked for Exxon and other companies in Peru, Libya, Nigeria, and Indonesia. For the Denton ordinance rewrite, his most notable contribution would turn out to be his proposal to include a welcome statement for the industry. He believed God had blessed Denton by storing this bounteous gift of natural gas under its soil.[22] One night, after another marathon meeting at City Hall, we talked together in the parking lot as I was attaching lights to my bike for the ride home. It was February 2012, a few days after MG's fourth birthday party at the fire station downtown. City Council had just passed a moratorium on new drilling permits while the task force did its work. I remember seeing Orion's belt through the fog of our breath as we spoke. At one point, John leaned back with his hands in the pockets of his jeans and said, "I don't understand why folks want to hide gas wells behind walls. A good gas well is a real thing of beauty. . . . It's a work of art!"

John would disagree with my then-budding conception of fracking as a real-world experiment. At most he might consider it something like a high school biology experiment that has been done a million times before. You know what the expected outcome is, and as long as you follow the book, getting the right result is automatic. Things will only go wrong if a lazy student—or a bad oil and gas operator—cuts corners. This was the rhetoric that dominated the task force: fracking is inherently safe, with a sterling track record that is only smudged here and there by a few bad apples.

Vicki, one of the nonindustry members of the task force, tried her best to offer an alternative framing of the issue. She pushed particularly hard on the precautionary idea of requiring comprehensive environmental impact assessments prior to permitting drilling. But after half a dozen meetings of spirited debate, the industry majority bloc settled on the silent treatment. Vicki's ideas mostly fell stillborn on an unreceptive panel.[23] Indeed, the task force did remarkably little to actually study the issue of fracking in Denton. They did not attempt to survey the public health literature. They did not call for any environmental studies. They didn't even bother to review the local economic impacts of fracking. There was no systematic cost/benefit assessment. As Vicki told me, everyone just assumed that 270 gas wells in the city must have a major positive economic impact. They also seemed to assume that local residents owned all the mineral rights. It wouldn't be until much later that I would learn how wrong those assumptions were.

Yet I always maintained that the biggest problem with the task force was not the industry majority but the behind-the-scenes bureaucracy.[24] In one of the early meetings of the task force, Mark Cunningham, who was Denton's planning director at the time, interrupted the conversation to tell them that they really only needed to supply general concepts; his office and the city attorneys would write the language of the ordinance. My sense was that the city bureaucrats still wanted to control the process and make the substantive decisions, which would have made the participatory democratic elements of the process just a facade atop a bureaucratic technocracy.

The task force wrapped up its work in March of 2012. The ordinance then went behind closed doors for six months to be vetted by lawyers and bureaucrats. During this time, in April, DAG hosted a mayoral candidate debate on the UNT campus between the incumbent, Mark Burroughs, and the challenger, Neil Durrance.[25] It was the only single-issue debate of the election season, and it drew the largest crowd, with people sitting in the aisles and standing along the back wall. I moderated the debate by selecting and reading questions from the audience to the candidates. Amber, who was hugely pregnant with Lulu then, volunteered to be the timekeeper, giving the candidates three minutes for each response.

There was one question I particularly liked: "Has fracking been good or bad for Denton?" Mr. Burroughs is tall with dark hair and has an imposing political presence, like Lincoln minus the top hat. He eventually eked out a win by carrying Southridge, one of the wealthiest and most Republican districts in Denton. When he stood up to answer the question he unequivocally said that fracking had been bad for the city. Mr. Durrance, who is much shorter and more portly, nodded in agreement, as did the entire crowd of two hundred people. When it was his turn to speak to the question, Mr. Durrance said, "Of course fracking has been bad for Denton," and then went on to explain how he would be tougher on the industry than Mayor Burroughs had been. At about the same time that President Barack Obama and Governor Mitt Romney were fighting over who would best promote domestic oil and gas production, the candidates for mayor of Denton were fighting over who would best oppose it.

Indeed, the entire ordinance rewrite was almost exclusively framed as a question of how strict the city could be on fracking. Almost everyone on City Council let it be known in one way or another that they didn't think fracking belonged in the city. Every citizen who took the time to speak their piece for three minutes at the end of task force meetings was opposed to fracking. The only ones who spoke in favor of it were a few young men and women with shirts from Clean Resources (another industry public relations group) who sheepishly admitted, when pressed by Vicki, that they had driven up to Denton from Fort Worth. They were likely being paid to show up. It was astroturf politics at its lamest.

One of the first people I spoke with as the head of DAG was the mayor pro tem, Pete Kamp, a fiery woman with a diminutive stature and a sugary southern drawl. I began by asking her what she thought Denton should do about fracking. Her response was telling. "Oh, honey," she said, "we all know what we *should* do about fracking. The question is what, legally, *can* we do about it." We wanted to be strict on the industry. That wasn't the question. The real fight was about how strict we could be within the limits of the law and, by extension, about whether the law itself was ethical.

When the city staff and attorneys finally released a new draft of the ordinance in September 2012, DAG members saw right away that it was a disaster. Not only were most of our recommendations not included, but the changes that were made were written in such dense legalese that no one could understand them. They seemed, however, to offer a convoluted way for gas well operators to bypass the ordinance entirely. For six months, policy engineers had been renovating the system behind closed doors, and their main value seemed to be not the protection of the public but rather the avoidance of an industry lawsuit.

Here's a sample of what I mean, taken from the revised ordinance:

> Unless the City determines that an exemption provided under Texas Local Government Code, Section 245.004 or successor statute applies to an amendment to the standards and procedures in DDC Subchapters 35.16, 35.22 and 35.23, or that Texas Local Government Code, Chapter 245 otherwise is inapplicable to permits for gas well drilling and production, such standards or procedures, except to the extent necessary to give effect to this subsection F, do not apply to the authorizations identified in subsection 35.22.4.A, if, on the effective date of such amendatory ordinance, the following circumstances existed. . . .[26]

It then lists seven circumstances. I still don't really understand what this means. Some theorists think that modern democracy is "a form of collaboration of ignorant people and experts." [27] Really, "collaboration" is being too generous. The experts basically run the show.

Once the revisions were made public, our policy window started to close fast. In October, we issued another report that literally cut-and-pasted existing language from other ordinances from cities on the Barnett that we thought Denton should adopt. In November and December, City Council vetted the draft ordinance both behind closed doors and in open sessions. New drafts emerged in a confusing fashion. Often the changes were not tracked, so as the document approached sixty pages in length we had a hard time seeing exactly what was being proposed.

A year earlier, I had felt so confident that our ideas would be implemented. I was now feeling frustrated. In December, I had a meeting with Darren Groth, the head of the new Gas Well Inspections Division. We marched through all the recommendations that DAG advocated, and he offered a point-by-point account for why they were not feasible. This one would constitute a regulatory taking. That one would trigger a preemption lawsuit. The other one was prevented by vested rights, state laws that grandfather existing projects under older rules. In short, his message was that the city's hands were tied. I only found out after the final vote on the ordinance that the industry was pressuring the city hard behind the scenes. Their attorneys were sending detailed comments on the draft ordinance, claiming time and again that municipalities lacked the authority to regulate the industry. Around the same time, I even drove out of town to visit the regional headquarters of one of Denton's main gas well operators (no operators are actually based in Denton), hoping to find some way to work around these barriers. Their community liaison and attorney were plenty nice to me, but each suggestion I offered was met with an immoveable wall as they said, in effect, "there is no need for us to do that and we are not required to do that."

Shortly after these meetings, Ginger Simonson called me up. Ginger, one of the members of DAG, was a retired US Army lieutenant colonel trained in logistics management and familiar with counterinsurgency tactics. Like Sharon and Cathy, Ginger was pulled into the fracking fight through personal experience. After her hometown of Flower Mound, just south of Denton, approved a gas collection facility, she helped the city craft a stronger ordinance.

When she called me, Ginger was driving down I-35, and her military training was in full effect as she simultaneously barked at me and the other cars, which were all, apparently, being driven by morons. Her staccato speech was drenched in impatience: "They are widening the Panama Canal. . . . China is investing in supertankers. . . . Use your signal!" I wasn't sure why she had called, but I tried to take notes. Soon it became clear that she was calling to dress me down: "The one place you have the power to influence the gas industry is the municipal level. And you guys in Denton are blowing it." Then she spoke to me as the leader of the Denton campaign: "Don't be Chamberlain. Be Churchill. Tap your inner Churchill!" She thought I was sliding down the road of appeasement, the way British prime minister Neville Chamberlain did with Hitler, as our policy recommendations and our precautionary ideals were systematically tossed aside for being impractical.

ON JANUARY 15, 2013, City Council was slated to vote on the revised drilling and production ordinance. Amber and I hosted a DAG strategizing meeting in our home about a week earlier. We also invited members of Denton off Fossil Fuels (DoFF), a group opposed to fracking and largely run by a collection of thoughtful college students and young citizens, and Occupy Denton. We all sat in the living room while MG ran around in a Batman cape (that was his superhero phase) and Lulu pulled herself up on the couch and coffee table trying to grab people's drinks.

The question was what to do at the City Council meeting. We all agreed the new ordinance did not go far enough to protect public health and safety and, thus, did not achieve the goal of compatibility. The setback distance for wells from homes was still 1,000 feet, though we had asked for 1,500. Compressor stations—loud and very polluting engines that pump gas down pipelines—were still allowed in town. Pits to store fracking fluids were still allowed. There was no requirement for air and water monitoring or a comprehensive environmental impact or waste management plan. Venting and flaring were still allowed. The use of low-toxicity drilling fluids and low bleed valves was not required.

Many members of DoFF had come to see the entire exercise as a giant smoke screen. If you cut through all the nuances and legalese, they thought, a simple conclusion was evident: we had to ban fracking. It was flat-out wrong. At that time, several cities in New York and one in Colorado had banned fracking, so the feeling was in the air that we could push the envelope much further. I could tell that many of the DoFF members, like Ricardo Orion, respected the work of DAG but thought that we had been lulled into the immoral sleep of moderation by all the complexities. Somehow I had drifted off course and got stuck in the weeds, losing sight of the big picture and instead advising City Council, for example, to modify the phrasing of Section 35.22.5.A.2.p.i of the ordinance. Where was my moral compass?

I respected that view. Ricardo and other members of DoFF made great points that night. They reminded me of the idealism I felt as a college student. Back then, I learned about the divide in environmentalism between the moderate Gifford Pinchot and the radical preservationist John Muir. Pinchot always followed something like the compatibility strategy, seeking ways to conserve some trees and also allow logging and dams. After all, we have to get our wood and irrigation water from somewhere. Muir, however, resonated on a deeper spiritual level. He refused the compromise intrinsic to the compatibility strategy. The forest is holy. The valley is sacred. Pinchot, I was sure, was a feckless weasel—paying lip service to these ideals while making decisions that undermined them.

But now I had become Pinchot as I responded to the DoFF group. I remember saying that we had to accept some compromise. A ban was too radical and could lead to harmful lawsuits. I told them, "I won't sacrifice the city of Denton on the altar of your idealism."

Despite the differences between DAG and DoFF, we left that meeting with the consensus that at the public hearing on the fifteenth we would all wear green shirts and stickers that read "Don't Frack with Denton." Several people also planned to speak longer than their three-minute allotment in a show of civil disobedience. Some of us didn't want to do that, but we did pledge to stand in solidarity in the audience every time a DAG or DoFF member spoke.

Then things got murky. In a sudden move in a work session minutes before the public hearing, City Council indicated they would increase the setback distance to 1,200 feet and implement an air and water monitoring program. I thought this was a big breakthrough, so I quickly moderated the tone of my own planned remarks to encourage more efforts like that. Vicki and Tom, the nonindustry voices on the task force, also took a more moderate tack. They had written a "minority report," dissenting from many of the recommendations of their own task force. The night of the vote, they detailed their own, minority, recommendations for the ordinance, hoping to nudge the council a bit further down the road of protections for public health and safety.

But for others it was too little too late. Of the thirty people that spoke from the standing room only crowd at the public hearing, thirteen were escorted out. They spoke longer than their allotted time, and they ignored pleas from Mayor Burroughs to stop. So the friendly Officer Brandon Hobon would approach the podium in the front and gently walk them out of the room, holding them by the arm. Each time a person was escorted out, the remaining crowd applauded. Cathy and Sharon were among those escorted out. Officer Hobon was later asked if he thought the situation would get out of hand, and he said, "No. I know these people. They've been coming here for four years."[28] He and Cathy, in particular, had formed a cordial bond over the years that she had been making three-minute speeches about fracking to City Council and the task force.

After a while, I stopped standing in solidarity with all the speakers. Many of them, I thought, crossed a line by demonizing City Council. I could tell by the faces of the council members that the disrespect did us no good—they looked embittered, tired of getting beaten up by both sides, and ready for it to all be over with. The final vote happened after midnight. The ordinance was approved 6 to 1, with the lone dissenting vote from James King, who likely thought it was too restrictive. I biked home exhausted through the dark and cold neighborhoods of Denton, thinking I would maybe write one more blog post but that I was done. No more DAG. No more fracking politics. Things were far from perfect, but they were slightly better, and we had done our best.

DAG went into hibernation for several months. But in early September 2013, just as I was getting into my classes for the new school year, I found myself sitting with City Council member Dalton Gregory at Big Mike's coffee shop near the UNT campus. A few days earlier, Cathy had sent me a panicked e-mail saying that EagleRidge Energy was drilling three wells on two sites on the corner of Bonnie Brae and Vintage Boulevard, a few miles south of the purple park. Sandwiched between the sites, one on the north and one on the south, was a neighborhood known as the Meadows at Hickory Creek. Some homes were closer than 200 feet from the sites. It was the very kind of neighborhood industrialization we sought to prevent with the ordinance. Yet here it was happening, despite the new requirement for a 1,200-foot setback distance.

I asked Dalton, who has a round, smiling face topped with a cloud of white hair, to explain to me what happened. He has an aura of patient wisdom that comes from serving as a principal in Denton's school district for years before retiring and joining City Council. Over coffee at Big Mike's he explained that our ordinance would only apply to situations where new gas wells come to existing neighborhoods—but most of the future fracking in Denton would happen in the reverse situation, where new neighborhoods come to existing gas wells. (Recall that Denton is growing fast, mostly to the west, where fracking has already pocked the land with hundreds of wells.) The ordinance presumes that homebuyers in that situation are knowingly consenting to the presence of the gas wells when they buy their home. So the "reverse setback" distance (when new homes come to old wells) was significantly reduced. That would allow developers to maximize their profits and allow fracking operators to continue their work. It's true that we had some inkling about the issue of reverse setbacks during the ordinance revision process, but it had never received much attention either from the council or the task force.

Again, that's probably because everyone assumed that people who buy homes near existing gas well sites will be adequately informed about the risks. However, no one in the Meadows neighborhood on South Bonnie Brae knew about the gas wells. The residents there also didn't know that EagleRidge was about to not only turn the two existing wells from vertical holes to horizontal ones but also add a new well to the northern site (we'll come back to this issue in chapter 5). None of this surprised Phyllis Wolper, a real estate agent and member of DAG. Phyllis had seen the ways fracking could negatively impact neighborhoods. Throughout our DAG meetings and forums she stressed the problem of disclosure and the fact that it wasn't uncommon for developers and real estate agents to mislead people. Even if they didn't actively lie, they often just simply didn't know enough about gas well development to inform potential homebuyers of the risks involved. As it turns out, we all should have listened more attentively to Phyllis.

There was also another major problem: vested rights. As Dalton explained it to me, according to Texas law, once a gas well site has been platted or permitted (even if there are no actual gas wells there yet), it is vested under the laws in place at that time. This is where things get metaphysical, because the "new" well on the northern pad site actually would not be "new" at all. It is part of an ongoing project that was first permitted in 2003. So, in a way, it has been in existence all along. That meant someone could move into a home prior to any gas wells physically being located nearby, but if that land was platted for gas well development a "new" well could be fracked 200 feet away, because it wasn't really new at all. It already had some form of existence, even if it was in a kind of legal limbo. To make matters worse, as Cathy and Sharon were to later learn after hours spent poring over dusty old files at the fire department, over 11,000 acres of the city's land had already been platted for gas well development. The industry was claiming this gave them permission to frack as many wells as they wanted "in perpetuity" under older regulations, *not* the new ordinance.[29] There was also some talk about vested rights during the ordinance revision, but for the most part it was the elephant in the room that people generally managed to ignore.

Dalton put it succinctly: "With our new ordinance, we only closed the barn doors after the cows had already got out." As Denton came to learn more about the dangers of fracking, it would be unable to do anything about them, haunted as it was by the ghosts of policies past. "Yup," Dalton said, "we're like a country dog in the city. There's no way out. If we do nothing, we'll get run over with more fracking. But if we move to stop the industry, they'll rain hellfire from their deep pockets." I could tell he was frustrated by the whole situation. As we left, he said, "The toughest thing about being on City Council is wanting something real bad and having your attorneys tell you that you just can't have it. Now, you can always get a lawyer to tell you what you want to hear ... but sometimes you have to listen to the unpalatable message."

A few days later, Cathy and I hitched a ride with Rhonda Love from the UNT campus to the Meadows neighborhood. Rhonda, a member of DAG and a retired professor of public health, moved to Denton in 2009. Although she grew up in Texas, she had been teaching for several years in Toronto. Shortly after she arrived in Denton, she discovered that several gas well sites were located near her home in the Southridge neighborhood. As part of her contribution to DAG, Rhonda put together a report for City Council that surveyed the existing public health literature on fracking.[30]

On that warm September day, Rhonda was looking every bit the Texan as she drove Cathy and me in her big maroon sedan, wearing big shades with a big hat atop her silver hair. One of the residents in the Meadows area, Sandy Mattox, had worked with Sharon to design flyers about the drilling. Sandy had been getting headaches and asthma-like symptoms ever since the rigs started coughing up diesel fumes day and night. In fact, it was Sandy who first sounded the alarm about fracking in that neighborhood. She had initially called the *DRC* asking them to write about two drilling rigs going up in a neighborhood. The paper told her it didn't really constitute much of a story, because fracking was so commonplace. Sandy thought to herself, "To hell it isn't a story!" and called up Cathy. The flyers Sandy made had the contact information for the Texas Commission on Environmental Quality, City Council members, and the Denton Gas Well Inspections Division. Cathy, Rhonda, and I had volunteered to go door to door to hand them out.

The Meadows is a nice neighborhood with new homes priced within reach of young families. It was about three in the afternoon when we were there. At one point, on Shagbark Drive, I paused to take a picture of the rig on the southern site. As I pulled out my phone, a yellow school bus rounded the corner and stopped no more than 150

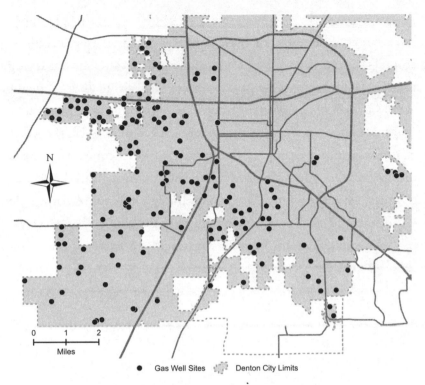

Gas Well Sites in Denton's City Limits

The city of Denton encompasses about 60,000 acres, sprawling irregularly to the north of Dallas due to a legacy of patchwork land acquisitions through annexation. In the city limits of Denton there are over 280 gas wells, which are distributed across roughly 150 separate sites (some sites have multiple gas wells). If you crammed all of Texas's gas wells into the 7 percent of its land area that is developed, Denton would still have three times as many wells as the average city. *Courtesy Zachary Coomes.*

feet from the rig, where a bunch of young moms were waiting anxiously. Many of them took their child's hand and walked quickly straight to their homes. I handed one mom a flyer as she passed by, and she thanked me over the screech of the rigs around us, adding as she kept walking home, "I can't believe they allowed this to happen." I don't know who she meant by "they," but I felt a twinge of guilt. How *did* we let this happen?

As I watched the children get off the school bus, I thought about how that could be Lulu or MG. Little Lulu had to use a nebulizer around that time, and I always suspected it was because she was trying to grow a pair of lungs in a place with such bad air quality. I thought about how angry and worried I would be if my kids were exposed to that noise and those fumes day and night for months, with the worst of it, the actual hydraulic fracturing stage, yet to come. Maybe the members of DoFF had it pegged: when you cut through all the details, this was just plain wrong. We'd gone through years of rewriting the ordinance, and still this was happening to these families—and it could happen again and again. Maybe something else, something more radical, had to be done.

SHALE VOYAGES

I T HAD RAINED ALL WEEK, AND THE OCTOBER SKY was pregnant with more rain. The wind was cool on my fingers and the road was wet under my tires as I biked south across what MG called "the busy road" and down Avenue C toward campus. It had been a month since I snapped that picture of the school bus in the shadow of a drilling rig at the Meadows neighborhood. While they continued to drill there, EagleRidge was hydraulically fracturing two wells on a pad site across Bonnie Brae from the UNT campus. That site was about halfway between the Meadows to the south and the purple park to the north. When the rigs were done drilling at the Meadows, the hydraulic fracturing crew currently at UNT would move south to frack those wells.

From my visits to the Meadows in the past month, I had gotten some sense of the drilling side of the operation. But I wanted to see the hydraulic fracturing that comes after drilling and that was about to visit the Meadows. So on that gray morning I rode my bike past my office on the north side of campus to the frack site on the southern edge by our newly constructed football stadium.

It was a dark time for me. I had been losing sleep, agonizing about what, if anything, DAG could do to help resolve the lingering fracking problem in Denton. For two years, we had marched through a process of meetings and reports as we revised the gas well ordinance. It was the epitome of order, of sanity. But the result was insane. Thanks to the predominance of the mineral estate, vested rights laws, and Texas's lust for mineral wealth, we had a heavy industrial activity permitted in the heart of neighborhoods.[1] Young mothers—Maile Bush, Kelli Higgins, Alyse Ogletree, and more—felt helpless, trapped in their homes, worried about their children. How was this rational? The thought of going back to City Hall and back to the drawing board was exhausting. What were we going to do, revise the revised ordinance? Everyone was burned out, and it seemed to be futile even if we could work up the energy.

When I took that bike ride to the UNT frack site, DAG was hanging by a thread, sustained only by the occasional meeting at Rhonda's house.[2] I had my usual reclining chair by the fireplace; Cathy and Carol were on the couch; Ed had another chair, fittingly, by the drum set that Rhonda was learning to play. Rhonda sat next to Ed recording the minutes of our meetings, but we didn't have much to report. We just sipped coffee and ran round and round the question: What to do?

That question was weighing on our minds two weeks earlier when we did a presentation in the community room at Fire Station #7 next to the Meadows. Cathy and Sharon told the roomful of anxious residents, including young children and pregnant women, about their own experiences with fracking near their homes. They also gave the people some information about how to conduct air quality tests and track and report health symptoms.

In my presentation, I showed how no one in that neighborhood owned mineral rights (which were held by four companies and trusts, none based in Denton). I had learned this information after several weeks of feverish searching through the database maintained by the

Denton Central Appraisal District. I had started blogging my initial findings about individual wells that summer. This got the attention of UNT geography professor Matthew Fry, who patiently helped a befuddled philosopher figure out how to handle all those data. I also explained to the Meadows residents in my talk at the fire station how, thanks to preexisting gas well plats (some of which were over two hundred acres in size), there could be dozens of neighborhoods in Denton like the Meadows with fracking as close as 200 feet from homes. I didn't say this to those families in attendance, but it seemed that DAG had a bleak future ahead of itself—years spent going to community rooms like this one to tell people their neighborhood just became an industrial zone.

Those were the thoughts going through my head, dark as the swirling clouds above, as I biked through campus. How bad was hydraulic fracturing, really? That's what I needed to know. But there was no objective answer to that question. It was maddening. The science was so sparse, and what was there was often contradictory.

I was ashamed to realize that although I had studied hydraulic fracturing for the past two years, with a few exceptions I had not really experienced it. There was one time on a hot Saturday morning in July when I had gone for a jog while MG was at tae kwon do practice. I was down near Denia Park, south of campus, when I heard a loud noise that sounded like someone was revving the engines on a 767 airplane. I jogged toward the sound until I was about a thousand feet away. It was coming from a gas well, but I couldn't get a good look at it. All I could see was a huge tank and a semi-truck. I tried to get closer, but there was an overpowering briny odor. I started feeling nauseated, like something sour was licking through my intestines and settling into my kidneys. I panicked and turned around. I remember feeling sick the rest of the morning, but I didn't want to believe it was from chemical exposure. I mean, maybe it was just the heat. I still don't know exactly what was going on, but I am pretty sure that I ran into the

"flowback" that happens for a day or so after a well has been hydraulically fractured and the chemicals come rushing back up with the liberated methane gas.

My only other real experience came about a year earlier, in November 2012, when I toured a "frac job" about two miles west of the UNT
campus with Mark Grawe, the head of EagleRidge Energy. Because
Mark's company had bought up lots of old gas wells closer to the core
of the city, he found himself in the middle of the fracking fight. For
a while, we forged a fairly cordial relationship. Cathy was always sus-
picious of this, but Kevin had asked me to see if Mark and I couldn't
craft some common policy proposals to forward to City Council as
they rewrote the ordinance. With a high forehead and a broad grin,
Mark could be John Elway's brother. Like the typical Texas oil and
gas man, he often wears jeans and cowboy boots and carries himself
with an air of self-assurance. It was bitterly cold the day I met him on
a fracking site, so I quickly jumped off my bike and into the heat of
his truck. Seen from behind the windshield, the hydraulic fracturing
operation had a muffled quietude that offered a comforting illusion, as
if what was happening was on a muted television.

When I arrived at the frack site near the football stadium, I realized it would be much harder to get a good look at it. By then, I was
no longer on speaking terms with Mark (I broke off our relationship
after seeing what his company was doing to the Meadows neighborhood). So I wouldn't be able to get past the security trailer along the
temporary dirt road being used to bring in truckloads of sand and
chemicals. The site was a few hundred feet off Bonnie Brae, and there
was the typical 12-foot beige sound barrier around three sides. But I
could see that the fourth side, which faced the highway on the opposite end of the field, was wide open. I'd have to go over I-35W, park
my bike off the frontage road, and play a real-world game of Frogger
across the highway.

When I got to the bottom of the hill leading up to the highway, I tried to park my bike, but the kickstand sank into the mud like a hot knife through butter, and it crashed with a pathetic rattle from its little bell. I was wearing my nice shoes because I had to teach later that day, but after slipping on the hill, muddy from recent rain, and planting my knee on what appeared to be a patch of poison ivy, I started to regret my choice of footgear. When I finally crested the hill, I slung my leg over the guardrail and waited for a break in the traffic. Cars and trucks zoomed past on the two lanes heading south on I-35W toward Fort Worth. I'd have to negotiate that side of the highway and then cross a grassy median before traversing the lanes of traffic heading north toward Oklahoma.

When my chance arrived, I seized the moment, sprinting as best as I could in my wet loafers across the concrete. My momentum carried me down into the median. The rain had turned it into a swamp, and as I crossed the lowest point, the mud stole my right shoe. I was left with one shoe and one dress sock, looking up at the speeding cars heading in opposite directions on both sides of me.

DURING THE SEVENTEENTH-CENTURY religious wars in Europe, Thomas Hobbes diagnosed a condition of "double vision." What is truth for one faction is heresy for the other, and vice versa. What's accepted as reality depends on incommensurable worldviews. Something similar could be said about the fracking debate.[3] Both sides seem to have an unshakable faith that fracking is either good or evil. Both are convinced the evidence confirms their view. As City Council member Dalton Gregory would later point out, the clash about fracking is similar to the one about alcohol. Quoting a Mississippi politician from the 1970s, Dalton wrote on his blog that "whiskey" can mean "the devil's brew . . . that destroys the home," or it can mean "the oil of conversation" that "magnifies joy."[4] Depending on how you choose to frame fracking—focus, say, on threats to water or benefits to economies—you could tell similarly contradictory stories about it too.[5]

Is the divide over fracking the same old partisan one between the left and the right? There's certainly some of that going on. But that's not the whole story, because the warring camps don't always map atop our red and blue political terrain.[6] Consider the rift fracking has caused in the Democratic Party in Colorado, where it sparked high-stakes brinkmanship between Governor John Hickenlooper and US Representative Jared Polis, both Democrats.[7] At the federal level, both Obama and Romney embraced fracking in the 2012 presidential race. Calvin Tillman, former Mayor of Dish, and several people in the Meadows, like Debbie Ingram, are conservatives but oppose fracking to one degree or another.

Maybe it's just proximity that determines the divide: the closer you live to fracking (especially if you don't own the mineral rights), the less you like it. There's doubtlessly much truth to that. But the more I read about the issue and the more I talked to the people of Denton—a mix of redneck conservatives, liberal college hippies, and much more, some living right next to wells and others miles away—the more I suspected something deeper was at work: the fracking debate is driven to some significant degree by different worldviews—as I see it, between pre-cautionaries and proactionaries.

To get a quick sense of this divide, recall the different reactions that President Carter and wildcatter George P. Mitchell had to the news in the 1970s and 1980s that domestic oil and gas production was dwindling. Same news, opposite interpretations. Carter took it as a sign that humans were exhausting a finite resource; we needed to rein ourselves in. Mitchell took it as a sign that we needed to push further—to tap into unlimited human ingenuity to unlock more resources. Carter had a precautionary view of nature as bounded and fragile and of humanity's place as circumscribed. Mitchell had a proactionary view of nature and humanity as malleable and full of potential yet to be realized. One is a world of finitude, caution, and proper order. The other is a world of risk taking and boundless possibilities.

This divide is most clearly visible in debates about biotechnology. Opposition to doping in sports, for example, unites left egalitarianism (it's not fair) and right cultural conservatism (it's an affront to human dignity). They form a precautionary alliance around the central theme of risk aversion: it's best to not fiddle with the established order. It was this precautionary disposition that the transhumanist Max More attacked with his idea of proaction. Why would we put up with the arbitrary limits imposed on us by a blind nature? We can take charge of our own evolution to sense and think and do things that are unimaginable today. This forward-looking vision of human freedom pulls together left-leaning progressives and right-leaning libertarians. In this way, biotech debates give the left-right political landscape a 90-degree twist to down (precaution) and up (proaction).[8] Precautionaries look down to our roots in the animal kingdom. Proactionaries look up to our future in the stars.

By the time of Carter's doomsday speech, the precautionary worldview was in ascendance, having displaced the postwar techno-hubris of "better living through chemistry." I absorbed that worldview as I fretted about sprawl and smog while growing up in Colorado. I can remember listening to my parents talk with their friends about the biologist Paul Ehrlich's 1968 book *The Population Bomb*. Ehrlich was the leading environmental voice at the time, achieving such popularity that he made frequent appearances on *The Tonight Show* with Johnny Carson (which was on after my bedtime). Ehrlich epitomized the precautionary worldview about the dangers of human overreach on a finite planet. The opening lines of *The Population Bomb* read, "The battle to feed all of humanity is over. In the 1970s and 1980s hundreds of millions of people will starve to death."[9]

The libertarian economist Julian Simon started out his career in the 1960s also believing in zero population growth. The Malthusian logic seemed unimpeachable: there's just so much planet to go around. But after success in the mail order business (and coming up with the idea to pay people to get off oversold flights), he started to look at the data behind human population growth and resource consumption. In short order, he radically changed his view to the belief that there are in fact too *few* humans. His 1981 book *The Ultimate Resource* argued that technological development and wealth generation are actually making natural resources *less* scarce. This is another kind of common-sense belief capable of rivaling the precautionary view of finitude: despite having consumed resources for centuries, we live in an age of unprecedented abundance. Simon became the leading cornucopian voice fighting Carter, Ehrlich, and all the doomsayers. Contrary to their litany of environmental fears, humanity was in fact better off than it had ever been before, and as long as enough human minds were free to innovate, progress will continue indefinitely.[10]

Simon couldn't stomach Ehrlich's popularity. He fumed as he witnessed a "juggernaut of environmentalist hysteria" take over the nation.[11] Simon was collecting reams of data to support his view. He was like the early Rocky, an unknown boxer training to take on the reigning heavyweight champion. Simon knew that he could defeat Ehrlich, and in 1980, he called him up and challenged him to a fight.

Simon proposed a wager. He bet that any five commodity metals, of Ehrlich's choosing, would all be cheaper within any time frame longer than a year. Working with his colleagues (including John Holdren, later to become President Obama's science and technology adviser), Ehrlich picked five metals that he was sure would increase in price: copper, chromium, nickel, tin, and tungsten. He also chose a decade as the length of time for the wager. Simon and Ehrlich each bought $200 worth of each metal for a total of $1,000. When 1990 rolled around, the price for all five commodities had decreased—they were, in Simon's terms, *less* scarce despite the fact that people had been consuming them for the past ten years. Simon won the bet. Ehrlich mailed him a check for $576.07, the difference in the prices.[12]

Simon made his wager two years before Mitchell tapped the Barnett, and he died in 1998, the same year Mitchell cracked the shale code with the S. H. Griffin well outside of Denton. Simon didn't live to see the shale revolution take off, but he would not have been surprised that we now have a glut of oil and natural gas after years of thinking domestic supplies were dwindling. Ehrlich and Carter may have thought that humans were running into the inevitable wall of a finite Earth. Simon believed we were on a never-ending escalator of progress.

When asked about his losing gamble, Ehrlich said Simon was like the man who jumps off of the Empire State Building and comments that all looks rosy as he passes the tenth floor. Eventually he will go splat. But that "eventually" starts to wear thin after enough false predictions of disaster. It's the same cold comfort for every prophet of doom holding a cardboard sign long after the scheduled end of time.

Simon said he won because he had worked out the logic of scarcity in a free society. More people consuming more goods cause resources to become scarce. Now, all scarcity really means is that something has become more expensive. This causes problems in the short run, but problems are also opportunities. Higher prices create incentives for scientists, engineers, and entrepreneurs to either find replacements (as when fiber optics and plastics replace copper wire) or discover more reserves (as when Mitchell unlocked the kitchen rocks deep underground, making unconventional sources of oil and gas economically viable). As long as people operate in an open society, they will invent a solution, because human ingenuity is the ultimate resource.

The Barnett Shale is a great example. In 1996, scientists estimated it held 3 trillion cubic feet (Tcf) of gas. By 2013, some studies had moved that estimate up to 45 Tcf of remaining gas, and that's despite the fact that in the intervening years 13 Tcf had already been extracted.[13] Simon might point to this example and say, "See, technological

advance means that the Barnett has more gas today than ever before!"
This shows how Simon's wager with Ehrlich was just an instance of
our society's more encompassing technological wager: betting on the
success of future innovations to bail us out of problems created by
present innovations.

Today's leading proactionary voice in the field of energy politics is probably Daniel Yergin. Echoing Simon, Yergin concludes his
award-winning book *The Quest* by writing that we can be confident
we'll solve our energy problems because of "the increasing availability
of what may be the most important resource of all—human creativity.
. . . And that resource base is growing." Ultimately, the human search
for energy is "a quest that will never end." [14] The proactionary mentality is also alive and well at the Texas Railroad Commission, the main
oil and gas regulatory body for the state. In 2013, its chairman, Barry
Smitherman, told the Society for Exploration Geophysicists that there
is a "relatively boundless supply" of oil and gas worldwide.[15]

But won't burning that boundless supply of carbon doom the climate? Many scientists estimate that around 3,000 gigatons of proven
reserves of fossil fuels must stay underground if we are to have any
chance of averting a climate catastrophe.[16] How far can Simon's logic
take us? I mean, maybe we can eventually recycle all the fracking water,
capture all the flowback fumes, and even drill ten-mile-long horizontal wells to reduce surface impacts.[17] But we can't engineer our way out
of the carbon problem, can we? The proactionary response: of course
we can! We can capture and store carbon; we can even regulate the
planet's thermostat via geoengineering schemes like reflective mirrors
in outer space or artificial blooms of algae.[18] We can control planet
Earth: this is our civilization's ultimate technological gamble.

It's tempting to get caught up in the allure of the tech-fix, but
Ehrlich's precautionary view cannot be so easily dismissed. Simon said
scarcity, and thus price, could be reduced in two ways: find more of a
commodity or replace it with something else. But there's a third way to

lower the price of a commodity, namely, externalize costs. When costs are internalized, prices go up (meaning things become more scarce). For example, in an effort to internalize costs the City of Southlake, close to Denton, imposed so many rules on fracking that the cost became prohibitive, and Chesapeake Energy packed up their rigs and left town, grumbling about "unnecessary" regulations.[19] Indeed, this is another way to frame the work of DAG: with our recommendations, we were trying to internalize the public health and environmental costs of fracking.

Prices drop when regulations are evaded or reduced, because certain costs are not reflected in the price of the commodity. Ehrlich was most worried about "the capacity of our planet to buffer itself against human impacts."[20] The point is that this "resource" does not get a price, because it is not a commodity but rather part of the connective tissue that makes commodity production, and the absorption of waste, possible.

Ehrlich should have questioned the terms of Simon's wager, because they only asked *whether* the price had dropped, not *why*. So prices dropped over a decade, but someone with a precautionary disposition could still claim that was only because the true costs, and the long-term costs, were not being accounted for in the price.

In 2002, for example, an increase in tin mining (one of the five commodities in their wager) in Indonesia caused tin prices to plummet. But the mining destroyed 1.5 million acres of jungle, leaving craters filled with polluted water. Environmental and human health costs were not factored into the price of tin. The meager wages offered by the mining companies also kept the price low, with one miner reporting that he split ten dollars a day with two other miners. When Indonesia cracked down on unregulated mining, the price of tin went back up.[21]

Note that Simon's logic actually depends on problems that increase prices, because that is what spurs innovation and progress. All those plans to capture carbon and engineer the climate depend on carbon getting a price so that such schemes actually become cost-effective. Keeping what Ehrlich might call the "true costs" of carbon out of its market price hampers innovation. But it serves the interests of corporations whose present business model relies on rules that have established the atmosphere as a free place to dump unlimited carbon emissions.

AFTER DRILLING DOWN into the standing water with my arm, I finally liberated my shoe from the mud and sloshed it back onto my foot. I was determined to get my view of a hydraulic fracturing operation from higher ground. When there was a brief clearing in the traffic heading north on I-35W, I dashed to the other side, leaving wet footprints in my wake, and stepped over the knee-high metal guardrail.

Once on the other side of the highway, I could finally see down inside the walls surrounding the frack site, and that's when the full scale of the industrial operation finally sank in. The droning of all the truck-sized pump engines was so loud that it drowned out the sound of the traffic just behind me on the highway. Billows of silica sand were wafting toward the athletic fields. The northerly winds were blowing the thick diesel fumes to the south, but occasionally the winds would shift, bringing the plume west in my direction. When that happened, my lungs felt greasy and my head light. It was worse than an airplane tarmac.

The two wells, UNT #3H and #4H, were initially drilled in 2004 by Enexco. Originally, one of them was an older-style vertical well, and the other was one of the first directional drills in the area, running about 1,200 feet east to collect gas over a mile directly below the spot where UNT's new football stadium would be built some eight years later. By October of 2013 when I was there, EagleRidge had turned both wells horizontal. This time, they ran the lateral pipes parallel to

one another in a southeast direction. The Barnett Shale is like a giant
2x4, and its layers of sediment look and act just like the grain of wood.
The direction of the laterals is perpendicular to this geological grain so
that the underground explosions they ignite create fractures that run
with the grain, causing bigger cracks.[22]

Ironically, the laterals from these wells ran directly below UNT's
three community-scale wind turbines. UNT is known as the "Mean
Green," a moniker that originated from its smothering football defense
and has since been attributed to its legendary defensive lineman Joe
Greene. Shortly before I joined the faculty, UNT gave an environmental twist to this football identity with the slogan "We mean green,"
as in we are committed to sustainability. The wind turbines were the
shining symbols of that commitment. But now UNT was also profiting from fracking, because although the wells are not on UNT's
surface property, UNT owns 33 percent of the mineral rights pooled
together by the two wells.

In the nine years since Enexco first drilled the wells, technology
had improved to the point where the laterals were nearly three times
as long as the original directional drill. Better technology also meant
a larger payload. When the wells were completed, they both produced
record amounts of gas. In November, the month after my visit to the
site, UNT #4H coughed up more gas than it had produced in the past
sixty months combined.[23] You'd have to carpet the campus with over a
hundred community-scale wind turbines to match the energy output
from those two gas wells.[24]

That economic and technical success brought with it more intense
surface impacts and resource uses. From my perch above, I took out
my camera to record the bustling scene below. A dozen semi-trucks
equipped with giant pumps were connected to a latticework of black
tubes that pulsed and converged on the wellheads, which sat like spiders in the center of the web. The sand was being hauled in by a stream
of trucks that transferred their loads into three truck-sized hoppers,

where a man stood poking the sand with a pole as it swirled around his head. A white trailer toward the edge of the site acted as the brains of the operation. Inside, Weatherford engineers in red jumpsuits with a Texas flag patch on their shoulders monitored over a dozen computer screens, controlling injection rates and chemical mixtures. At this frack job, they would pump roughly 6 million gallons of water, 3 million pounds of sand, and 100,000 gallons of chemicals down *each* well over the course of about two weeks.[25]

Watching all that sand being trucked in was melancholy, because it had been ripped out of the same Wisconsin woods that Aldo Leopold romanticized in such a formative way for me in *A Sand County Almanac*. It takes twenty-five rail cars full of sand to frack an average well, and some estimate that number will double as operators continue to experiment with using more sand.[26] All of this demand is leaving deep scars on Leopold's home close to my alma mater of St. John's. In 2010, there were five sand mines and five processing plants in Wisconsin. Three years later, over a hundred new mines and plants had been permitted.[27]

As I watched the sand being mixed on-site, it made me think how surprised a geologist might be in the distant future to discover 300-million-year-old shale somehow reticulated with angel hairs of 30,000-year-old sand. The horizontal lengths of the wells a mile below campus would first be shot with a perforating gun to blast cracks into the well casing. Then the Weatherford crew would lower a pair of inflatable "straddle packers" into the pipe to isolate a section. Next they'd start pumping the mixture of frack fluids and sand into that segment. At first, the pressure would just counterbalance the in situ force of the rocks bearing down on the open space in the pipe with the weight of the earth above. But slowly they'd increase the pressure with some keystrokes at the computer to around 5,300 psi, about the same force exerted by the Jaws of Life used by emergency responders to cut through cars.

Then the jet-black rock would begin to fracture into a mosaic of cracks. The pressure would create a network of miniature pathways. Methane that had been imprisoned for millions of years would suddenly feel the near weightlessness of the surface tugging at it. Molecular diffusion would coax it out and up into the wellhead next to the football stadium, the national network of pipes, and, soon enough, the insatiable inferno of American consumption.

The hydraulic fracturing would take about two weeks, after which the wells would "flow back" as the chemicals, sand, and water—along with the methane—would come rushing back up. In the same process that happens when scuba divers get "the bends" from depressurizing too fast, those dissolved gases would volatilize at the surface, coming out of solution and rushing up into the air. EagleRidge claimed on their website that they practiced "green completions," a process that is supposed to capture the chemical emissions and keep them from entering the atmosphere. But during flowback, about a week after my visit, Calvin Tillman would use a special infrared camera to film a thick plume of emissions coming off the site.[28] Several people in the area complained of headaches around then, and one woman who worked near the football stadium told me she felt so sick she had to leave the office.

When I was there, they were still actively pumping things into the earth. In the foreground, I noticed two long metal pipes snaking through the undergrowth below me, bringing in freshwater from a nearby water well. The contractors would mix the sand and chemicals with this water on-site before pumping it down the holes. I skidded down the hill to get a closer look. The water had made the pipes cold to the touch, because it had just come from the aquifer over 200 feet below. I could hear it rushing through those enclosed artificial rivers, making a faint pinging sound on its way to a far deeper oblivion.

When I scrambled back up the highway embankment, I could see that I had attracted the attention of the contractors. One of them had grabbed his own camera and was filming me as I filmed them. I decided it was time to leave.

ABOUT A MONTH LATER, over the Thanksgiving holiday, Amber and I decided to take MG and Lulu to the Perot Museum of Nature and Science in Dallas. On the third floor of the museum, sandwiched between the dinosaurs above and the biology lab below, is the dimly lit, futuristic Tom Hunt Energy Hall. With barely a nod to solar panels and wind turbines, the Energy Hall is primarily a tribute to the technological marvel that is fracking. I had been hearing stories from other parents about the Shale Voyager exhibit at the museum. One young dad just shook his head, saying, "You gotta see it to believe it. It's, well . . . it's a fracking ride. I guess . . . With a robot. Sort of." I knew then that it was time for a family fracking trip.

The Perot (as it is commonly known) is a 170-foot-tall, walk-in cubist sculpture. The feathered contours of its light gray concrete skin suggest the pancaked layers of a sedimentary rock formation. It's as if the building were a cross-section of millions of years' worth of limestone deposits that had settled on the bottom of an ancient sea and were now suddenly uplifted into the heart of Big D. The escalator is enclosed in a glass tube that protrudes out of the building. The way it slashes downward through the wall is more than a little reminiscent of a drill pipe cutting through the rock on its way to a production zone several floors further down.

We took the train to the Perot, which made it the perfect opportunity for us to try out our new red wagon. Lulu, who was not yet two, was a miniature tornado who would sprint about chaotically anytime we were so foolish as to let her own two feet touch the ground in a public setting. MG, who was five, was far tamer, but had an unfortunate habit of going boneless just at those times when carrying him was sure to be the most inconvenient of options.

They liked the wagon, and Lulu would actually sit still in it as long as it was moving. Of course, as soon as we got it on the train she clambered out of there with an ungainly flailing of her limbs, screaming "choo choo!" and inadvertently poking MG's eye in the process. And there we were with one kid crying, another one darting away, and a giant wagon blocking the aisle. As I watched the other passengers roll their eyes, it dawned on me that we had become *that* family.

The train tracks run a few miles east of Pilot Knob, a protrusion of ironstone and sandstone ledges that, from a distance, resemble a ship's lookout. Pilot Knob was used as a lookout station by the Caddo Indians, Spanish and American pioneers, and, later, the Texas Rangers (of the law enforcement, not the baseball, variety). According to legend, it was also used by the notorious outlaw Sam Bass as a hiding place for the gold he stole from the trains of his day. Pilot Knob is on the 3,300-acre Hunter Ranch, which occupies much of southwest Denton and is owned by the Perot family. One facet of their $5 billion empire is Hillwood Oil & Gas, which operates sixteen gas wells in Denton that have contributed roughly $13 million to the Perot family fortune.[29]

On the train ride, MG looked up from the game he had been playing on my new smart phone, an early Christmas gift from Amber, who apparently felt that I needed to join the twenty-first century. He said he wanted to see the dinosaurs first. I told him I wanted to see the fracking exhibit. From the time he was two years old, MG had been involved with fracking politics. So he knew a little something about the issue, and on the train he asked me his standard question once again: "Daddy, are gas wells good or bad?"

I was always flummoxed by that question, so I told him honestly, "I don't really know . . . it's complicated and depends on the situation." I could sense he wasn't satisfied with that, so I looked out the window at the sprawling metroplex zipping by and added, "I guess I think gas wells are pretty much bad. I mean, they make lots of pollution. But we rely on them. So if we say they are bad, it's kinda like saying there's something not right with the way you and I live every day. I don't think we should be doing all this fracking, but that means we have to do lots of things differently."

At that point, I looked over at MG only to find that he had gone back to his game. The sound of angry birds being launched at pigs was a fitting commentary on the plight of the field philosopher.

When we got to the museum, I tried to wait patiently for the Shale Voyager as MG marveled at the dinosaurs and Lulu waved at people going up and down in the glass elevators. I had not made a big deal about this "ride," because I didn't want the kids to be swayed by what one reporter called the main message of the energy exhibits: "Fracking is fun." [30] I was both eager and apprehensive to see how the Energy Hall had blurred the line between public education and corporate indoctrination. After all, it was named for a petroleum industry magnate who sold his company for over $4 billion in 2008 to XTO Energy, which (no surprise) operates gas wells in Denton.

In the age of Google, museums are becoming less about information (you can access that on your smart phone) and more about creating a mood. And the general mood of the Energy Hall with its soothing blue lights in dark recesses was one of calm and comfort. Everything that comprises our modern existence depends on energy: our clothing, our phones, the big red plastic wagon, the chocolate-milk boxes in the fridge down in the cafeteria, and on and on. Whatever anxieties about this we might have had before walking into the Energy Hall—about scarcity, pollution, or climate change—we can now release with a big exhale. There will always be scientists and engineers working at the frontiers of knowledge and in the bowels of the earth to conquer nature for the good of mankind. Somebody will be there to figure it out. In other words, the proactionary faith in innovation is the museum's hidden curriculum.

From one corner of the hall, under a giant replica of a drill bit the size of a truck, MG had a good time turning all the wheels and valves on a gas wellhead. It was about eight feet high and painted red. It was an *actual* gas wellhead. But it was also fake, because it was stripped of its context—the mud, the pipes, the trucks, the chemicals, the noise. It's the same way a tiger in a zoo is not really the real thing. The beauty of museums for the oil and gas industry is you don't have to lie. You can have real facts about real equipment. But when those devices are polished clean and arrayed neatly atop some nice new carpet, it is still a fiction. It's just such a seductive fiction, because everything is so sanitized. What could be wrong with this shiny piece of machinery that my kid is hanging from like a monkey?

The oil and gas industry wants fracking to "take on the harmless aspect of the familiar," as DDT had done by the early 1960s.[31] When you target children, you normalize for them what probably should seem like insanity. It took over 20 million diesel truck trips to bring the gas wells on the Barnett Shale into production.[32] That's absurd. But if you remember learning stuff like that at a museum while you were pushing buttons and chasing your sister, then, hey, no big deal. It's always been that way. "Progress" means one generation becoming completely used to things that seemed crazy to their parents.

The industry gets the same effect of "museumifying" the truth with their animated commercials. They're making us familiar not with the messy reality I saw that gray day next to the UNT campus but with a carefully constructed fiction: those digital gas wells fueling the lives of perfect families where Mom is perpetually pulling a turkey out of the oven. They're giving us the Disney World version of fracking, because what could be more familiar than Main Street USA?

AFTER HANGING FROM gas wells and ogling the giant drill bit, we decided to ride on the Shale Voyager, which looks like a metallic bunker nestled in the side of a plastic mountain. A smiling young museum volunteer opened the door and led us into the "4-D Theater

Experience." The inside was a cylindrical theater with a domed roof. Curving around the front was a screen, and around the back were about twenty black rubber seats arranged in two rows. They were the kind of seats you'd find on those racecar games at arcades. There were more screens above the seats, and a giant projector on the ceiling was casting yet another circular image on the floor directly in the center of the space at the foot of the seats.

Lulu immediately started climbing all over the seats as a couple of other families trickled in behind us before the door closed. The volunteer told Amber to hold on to Lulu, because the fourth *D* in our 4-D experience was going to be vibrating seats. MG just sat there enchanted by all the screens, but Lulu did not want to be held and let everyone know that fact as loudly as she could. I started to think this might have been a mistake.

But it was too late. The lights dimmed, and an attractive and perky blonde on the front screen started telling us about the Barnett Shale. She also informed us that we were not in a theater. No, we were sitting inside a robot named Otto, and now Otto was hitching a ride on board a tanker truck heading out to a gas well site. The side screens depicted a lush summer landscape somewhere in North Texas zooming by as we sped down the highway. The seats were jiggling with enough gusto to indicate this road was in dire need of repairs. I was having a hard time concentrating as Lulu ramped up her efforts to escape Amber's clutches and managed to lunge into my arms, but I think it was at that point that Otto "shrank" to the size of a golf ball, a process that apparently energizes seats into a whole new level of vibrational ecstasy.

Then Otto went whooshing 7,000 feet down the borehole on the floor through the artificial depths of virtual rocks. The lady on the screen said reassuring things about the many layers of cement casing around the well to ensure leaks could never happen. At some point we took a jarring right turn through the digital shale, and we managed to liberate the imaginary gas just as Lulu liberated herself from my lap and started to career dangerously around Otto's shaking innards. We then rushed back up the borehole, now floating in a geyser of methane, while I chased Lulu. The lady on the screen said some inspiring and patriotic words about how great fracking is. Then the lights came back up and we staggered out of our robot host, having survived a voyage through the shale with Lulu, our rogue adventurer.

Before we went downstairs to get lunch, we stopped to watch two short cartoons about fracking that were playing on TV monitors in the Energy Hall just outside the Shale Voyager. As if the fracking ride hadn't been enough, these cartoons took the museum's cheerleading for the oil and gas industry to a whole new level.

The first video featured a voice speaking over a cartoon explanation of fracking and the Barnett Shale, which has "enough gas to fill 288,461 Dallas Cowboy football stadiums"! The medium of the cartoon abstracts fracking from its grimy reality. Toward the end of the cartoon, the voice said, "There has been some controversy when it comes to drilling," as a character held up a sign that reads "Not in My Back Yard!" They boiled all the complexities of DAG's efforts—and the efforts of so many other community groups—into one grumpy stick figure. The cartoon quickly brushed past this to note that despite these "opinions," the fracking method perfected on the Barnett has unlocked shale gas reserves across the country.

The screen then showed some smiling scientists in lab coats surrounded by smiling children as the voice told us that "making all this production of new energy possible will require a whole new crop of scientists and engineers . . . one of them might even be you!" The cartoon concluded with the cheers of children and a triumphant blast of trumpets. I looked carefully at MG and Lulu after watching the cartoon to see if I could note any effects. They had no comment and seemed thoroughly unimpressed with the poor production quality. They ran over to the screen playing the other cartoon.

This one was more obnoxious. The whole thing was a kind of honky-tonk song sung in a crooning baritone, Elvis Presley style, by a cowboy with a guitar, bolo tie, and big white hat. The cowboy sang about the "competence and skill" of the landmen (corporate representatives who arrange the gas leases) who "have to jump through hoops and deal with special interest groups."

The chorus was:

Urban drilling in the shale; you can't leave out one detail.
But take the time to get it right and there'll be gas tomorrow night!

At that point, the screen showed the word "GAS" lit up in sparkling rhinestones. The chorus continued:

That natural gas that's underground,
Just hit that pay zone and smiles are all around.

Then three happy houses appear atop the shale, grinning and bouncing up and down as if they're in a 1930s Mickey Mouse cartoon. Lulu thought it was hilarious. For the rest of the day, she'd ask me, "What d'ya say, Daddy?" and I'd squint my eyes in my best cowboy impression and try to sound like John Wayne. "If we get it right, there'll be gas tomorrow night." It cracked her up every time.

CHAPTER 4

FRANKENSTEIN VS. FAUST

FOR CHRISTMAS THAT YEAR, SANTA BROUGHT US a Kindle. It was ostensibly for Amber, but MG immediately appropriated it and became obsessed with Minecraft, which is sort of like a digital version of Legos except you can fly around chucking werewolf eggs into forests and building traps to catch the werewolves after they hatch. I mean, if you're into that sort of thing.

I wasn't sure what to make of the game—yes, it fostered creativity, but it was still a screen, and we liked to pretend that we had rules about "screen time." So we made a deal with MG that we would also download and read some books on the Kindle. Reading, in my mind, would not count as screen time even when it was on a screen. I knew this was probably just a twentieth-century hang-up that my kids will not have when they're my age.

MG's interest in werewolves made me think he would like Mary Shelley's *Frankenstein*, so we downloaded it and started to read a few pages every night. The book opens aboard a ship on the barren Arctic icescape, and the preface is a series of letters written by a young explorer, the captain of the ship, to his sister back home in England.

As they forged northward and the ice closed in, he says, the ship's crew saw a mysterious figure, more giant than man, guiding a dogsled over the horizon. The next morning, they welcomed aboard a ragged man, half dead from hunger and fatigue, who had also been traveling by dogsled and was now afloat on a slab of ice. It was Dr. Frankenstein. He was in pursuit of his murderous creation.

After some days of convalescence, Frankenstein asked the captain to tell him about his voyage. The young captain was overeager:

How gladly I would sacrifice my fortune, my existence, my every hope, to the furtherance of my enterprise. One man's life or death were but a small price to pay for the acquirement of the knowledge which I sought.

(Shelley's thick prose made for slow reading with MG. After a passage like this I'd need to define words like "furtherance" and explain that "acquirement" was not like his favorite mint gum.) As the captain spoke, a gloom began to shroud Frankenstein's face. He had been down that road. He had made that bargain for infinite knowledge, and it had brought nothing but ruin and misery. With a flood of emotion, Frankenstein burst out:

Unhappy man! Do you share my madness? Have you drank also of the intoxicating draught? Hear me—let me reveal my tale, and you will dash the cup from your lips!

Frankenstein proceeded to tell his mournful story. Even as a child, he was "embued with a fervent longing to penetrate the secrets of nature." (Translation: he was like Sid the Science Kid, who "wants to know everything about everything.") Frankenstein was tortured by the limits of human knowledge and anguished about ways to discover the philosopher's stone. At the university, Frankenstein became enamored of the image of the scientist as hero, as the wielder of "new and almost unlimited powers." He got drunk on the godlike potential of humanity and vowed to unravel "the deepest mysteries of creation."

When he reached the iconic moment in his story where he animates his creation, Frankenstein could see the captain was on the edge of his seat, itching to know the secret recipe for life. But Frankenstein would not share. He said "enough" and closed the book of knowledge:

> I will not lead you on, unguarded and ardent as I then was, to your destruction and infallible misery. Learn from me . . . how dangerous is the acquirement of knowledge, and how much happier that man is who believes his native town to be the world, than he who aspires to become greater than his nature will allow.

It was as I read these lines aloud to a sleeping MG that I realized Shelley's Dr. Frankenstein is the embodiment of the precautionary worldview. Know your limits and do not transgress them or you will regret it. Indeed, although it is often associated with liberals, precaution is actually a very conservative disposition.[1] It tends "to prefer the familiar to the unknown . . . the tried to the untried."[2]

The next day, as I watched MG hopelessly try to teach Lulu the finer points of Minecraft, I started thinking more about the opening of *Frankenstein*, set on the sea amid blocks of floating ice. I realized the whole thing is a warning, a cautionary tale. In fact, the word "monster" comes from Latin *monere*, the same root as "monitor," and it means "to warn." Precautionaries tend to see omens of impending evil everywhere, closing in like ice against the hull of a forlorn ship. It's no wonder they often have paranoid tendencies.

Precaution is most often talked about in terms of a principle, though, rather than a state of mind. "The precautionary principle" is a translation of the German *Vorsorgeprinzip*, which means "fore-caring principle." [3] At its core, then, it calls for foresight, preparation, and taking more things into account prior to acting. [4] If we commit to a technology and then wait for science to prove its guilt, it may be too late. Like Dr. Frankenstein, we could wind up lamenting our monsters *after* people have been harmed or killed or after the environment has been despoiled. When it came to fracking in Texas, we were just shooting from the hip, letting thousands of gas wells bloom with only the narrow intention of making money. Everything else—impacts on communities, air, water, seismicity, etc.—was boxed out of consideration the same way those cartoons and shiny museum exhibits box out the ugly undersides of fracking.

By the end of December, DAG's conversations at Rhonda's house had turned seriously toward the option of a fracking ban within the city limits. I was still skeptical, but I was beginning to question much of what I had written a year earlier in a *Slate* article that doubted the wisdom of municipal fracking bans. [5] My arguments from that time seemed less strong to me now, and they often didn't seem to apply to the situation as I had come to understand it in Denton. But still, wasn't it unphilosophical to cut the Gordian knot of such a complex issue with a simple "no, not here"? If I did this, how could I avoid being tarred with the brush of radicalism, dismissed for having plunged into the abyss of irrationality?

Then again, maybe to initiate a ban would be the most philosophical thing to do. Aren't philosophers supposed to question laws and conventions, not work within their parameters? And aren't we supposed to fight for the underdog? It also seemed to be a practical possibility. We had talked informally with some lawyers who suggested a ban on all drilling and production would not be legally defensible—but a ban targeted at the hydraulic fracturing stage, the most noxious part, could work. We had studied Denton's charter, which allowed for a citizens' initiative process whereby we could write our own ordinance, gather signatures, and put it up for a popular vote. It was a live option.

For Ed, who is a strong advocate of the precautionary principle, the ban was clearly the right choice. "There is a lot of uncertainty about all this, but we've seen enough here in Denton to know we should err on the side of caution," he would say from his chair next to Rhonda's drums. Ed would get on a roll, conjuring up a preacher-like pitch and cadence: "The consequences when things go wrong can be disastrous. Meanwhile, these guys in the oil and gas industry take a page from big tobacco's playbook and spread misinformation and doubt to keep stalling any effective regulations. Just look at what they said down at Fire Station #7 about how all the health complaints down there could be chalked up to allergy season. It's ludicrous!" He was pointing his finger accusatorily in the air as if industry executives were in the room with us. Cathy nodded in agreement on the couch, and Carol chimed in, "That's right, just look how long we've been working on this issue and we're no better off than when we started. When do we say enough is enough?"[6] I would fold my fingers under my chin in intense concentration as I listened to the group. It would be such a drastic step . . . but I kept thinking about that nightmare scene in the Meadows neighborhood with kids surrounded by fracking. How else were we going to prevent that from happening again and again across the city?

The precautionary principle had by then become a leading framework for many opposed to fracking. It is most visible in state-level moratoriums like the one in place for years in New York while it conducted thousands of pages' worth of studies. Eventually, the conclusion of those studies was that hydraulic fracturing should not occur in New York "until the science provides sufficient information" about the risks and whether they can be managed.[7] The precautionary principle can also be found in local bans and moratoriums enacted or proposed by several cities that seek to prohibit fracking "until it can be proven safe."[8] This shifts the burden onto the industry to prove that their products are safe prior to commercialization. The technology is guilty until proven innocent. It's something like the Hippocratic Oath of "First, do no harm."

Texas Sharon came to many of our meetings. Like Dr. Frankenstein, she has long been sounding the warning about the fracking monster. It was around this time that we were both invited by Dr. Rosemary Candelario to talk to her interpretive dance class at Texas Woman's University that was going to create a public performance related to fracking. Sharon had a slide show depicting the long prairie grasses around her old home in Wise County. She talked movingly about how the grass dances and how the seasons give it different moods. There was the soft baby grass of spring, the exuberant teenager grass blowing in the summer wind, and then the adult grass with amber seed heads nodding in the autumn sun. By the end of it, we had all come to deeply appreciate grass.

And then she showed a slide of a huge patch of gray, withered grass that had been burned by a spill of fracking chemicals. She nearly cried as she explained how she had gone back to that spot every year for the past decade only to find it still dead and barren. "What are we doing to this planet?!" she asked, her eyes almost wild with desperation, as a hush fell over the young dancers sitting on the floor in their knee socks and tights.

Like Sharon, Josh Fox, the director of the popular *Gasland* documentaries, has also been sounding the warning about fracking. There is an early scene in *Gasland* where he meditates at a stream where he grew up as a boy fishing and skipping rocks. He thinks about the meaning of place as he worries about how fracking might shatter a fragile but all-important feeling of being at home in the world. This scene is little remarked, but it is most revealing: he is seeking a rootedness in the earth, the pending loss of which is the key to understanding his fretful project and much of the global concern about fracking. The flaming faucets later in the film are so jarring because they represent a breakdown in the proper order—the world is disjointed, out of

.whack. Like Frankenstein's monster, the pieces have been arranged in an unholy and foreboding way.

I think Fox is like the ship captain in *Frankenstein* who heeded the warning and changed course *before* it was too late. Before he signed a $100,000 lease to allow fracking on his Pennsylvania farm, he visited the nearby town of Dimock, where he saw an industrialized landscape, sick children, and greasy gray water. He had a precautionary reaction: we're hubristically overreaching, and we are going to pay for it. His *Gasland* films are warnings. They are reminiscent of the line in Karl Marx's *Manifesto of the Communist Party* where he writes that bourgeois society has conjured up such gigantic technological systems that it has become "like the sorcerer who is no longer able to control the powers of the nether world whom he has called up by his spells." [9] Marx noted that capitalists will try anything to prevent a revolution of those being harmed by the monster. In *Gasland 2*, Josh exposes their main strategy, namely, buy the government. [10]

Here's the question: Who, or what, is the monster? Is it the creature or the creator? When reading *Frankenstein* with MG, I thought the answer was obvious: the monster is that foul thing, that abomination. But as we continued in the book, I found MG identifying far more with the creature than with Dr. Frankenstein. The creature is a lonely, abandoned child. MG would worry about how the monster was going to take care of himself. How would he eat and find shelter from the storms? He's really a pitiable patchwork of a soul who just needs a better dad.

Dr. Frankenstein, by contrast, behaves in a less-than-human fashion. He gives life to this poor brute and then runs away. He pursues what appears to be a noble vision, but his actions put a lie to all that. It was only a pissing contest with God. He carried out his plans without heart, he failed to care for his creation, and he irresponsibly left it alone to wreak havoc on the world. So who's the monster: creature or creator, the technologies themselves or the corporations that breathe life into them?

For some fracking opponents, like the environmental scientist and activist Sandra Steingraber, it's the creature. She sees the technological system itself as "inherently" and "irredeemably" dangerous, such that it would be a monster and a menace regardless of who was parenting and what rules they implemented. "Safe fracking," she argues, "is an oxymoron." [11]

Others put the blame on the corporations, and this is where interesting alliances start to form. John Siegmund is a good example. He was the oil and gas engineer on Denton's task force who thought gas wells were works of art and who proposed putting a welcome statement for the industry in our drilling ordinance.

For John, the technological creature is just in need of a better father, that is, engineers rather than capitalists. "Gas wells do not need to pollute," he said at one point from the dais at City Hall. They only pollute when the free market, with its principle of profit, trumps engineering with its principle of quality. [12] He was certain that the gas would be extracted. There was no stopping that. The only question was whether it would be done right. And for an engineer that means making sure the engineers are in charge. In this sense, he shared the typical precautionary distrust of capitalism. The invisible hand of the economic market is shortsighted and inefficient. If you are making an enduring piece of art, you are going to be more mindful than if you are making a cash cow to be milked until it's time to sell it to the next operator down the line.

But it's not just the technology and the capitalists involved in fracking. It's also the consumers—all of us who are complicit one way or another, because we use electricity, plastics, chemicals, and all the other goods that fracking helps provide. Marx predicted that communism would take over the world once the capitalists' monsters got totally out of control. Why was he wrong? It's because capitalism started exercising a new and insidious form of control. It got inside of us. We started adapting our wants and aspirations to its imperatives. We took on a false consciousness, becoming happily enslaved to a consumer culture that creates and satisfies false needs we confuse for the real deal.[13] As a result we became apathetic.

So one way to see the situation is to argue that we're not so much oppressed by a ruling class as we are lulled by pleasure. Seen from this angle, the fight Fox and others are waging against corporations is most painfully a fight against the needs and desires they have planted deep within us.

BACK WHEN I WAS still speaking with Mark Grawe, the head of EagleRidge Energy, he would occasionally swing by my office on campus to continue our dialogues. From his perspective all the unwarranted fear of fracking stemmed from its *perceived* newness (actually we've been fracking for decades, Mark constantly reminded me). People worry about fracking when they daily accept far greater risks without batting an eye, simply because those risks are familiar. "Thousands of Americans die in traffic accidents every year," Mark said to me at one point across my desk, "but no one is calling for a car ban." In Denton, the Peterbilt truck factory emits 270 tons of volatile organic compounds annually, but it doesn't draw any criticism, because folks are used to it. There's a 7-Eleven gas station right next to MG's school, but I wasn't protesting that.

For Mark, people oppose fracking only because they love the familiar, and this gives them a skewed sense of reality. This is why the industry paints opponents of fracking as irrational. They are as backward as the Flat Earth Society, stuck in outdated beliefs, unfairly demonizing what they don't understand, and unwilling to pull themselves into a new reality.[14] This framing depicts the precautionary worldview as a kind of psychological disorder. Opponents of technology are not seeing things objectively. They fixate, as Max More wrote, on "*perceptions of risk instead of examining the real risks.*" The legal scholar Cass Sunstein argues that the precautionary principle does not provide policy guidance, because it slips into self-contradiction (e.g., "precautionary" delays on developing new drugs can actually create more risks by keeping treatments from those who need them). Even worse, the precautionary principle is an expression of many of our cognitive biases. In short, it is really just irrationality in disguise.[15]

I could tell that Mark thought I too suffered from an irrational perception of fracking. In the way he patiently reassured me and reminded me of the risks I already tolerated, he was trying to get me to see things *objectively*. I always felt like Mark was trying to save me from slipping into hysteria. It was as if he were talking me down from the roof as I contemplated a plunge. What he said contributed to my lingering hesitations as I listened to Ed, Cathy, and Carol and contemplated the ban. Would it just be the result of an exaggerated fear, the way so many people once fretted about "test tube babies" when in vitro fertilization was first used? Although we never talked about it in these terms, I imagined that Mark saw me as part of the "antiscience left" like those who refuse vaccines for their children, because they are driven by baseless fear or at least a highly skewed risk assessment.

We did, however, talk a bit about nuclear energy.[16] Though it is widely perceived as the riskiest and deadliest form of energy, maybe it's actually one of the safest. Arguably, it has caused fewer deaths than solar panels, which require several toxic chemicals to manufacture. Nuclear power, the argument runs, is unfairly demonized simply because it has this otherworldly reputation, this uncanny feel about it. The task for Mark and his industry is to keep fracking from falling into the same emotional abyss.

As I thought more about it, though, I came to see that what they are really doing is trying to define the terms of what counts as a rational or objective assessment of a technology. Now, a nuclear meltdown or a dirty bomb could render a city uninhabitable for centuries. So too, a fracking accident could poison an entire aquifer. Those may be outside possibilities, but to include them in one's assessment of the technology is not irrational. It's just being cautious. Although there are more or less reasonable perceptions of risk, there is no such thing as More's notion of "the real risks." There is no objectivity that can scrub away human judgment when it comes to assessing whether something is safe enough or too risky. We are, as I said in the introduction, doomed to philosophize.

When we speak of danger, it is always from within an assumed scheme of familiarity that gives us our bearings. Mark wanted me to see that our scheme of familiarity is a moving target. In Gram Deal's time, television was exotic, but now it's commonplace. When I graduated from St. John's, e-mail was still novel. Now I skim dozens of e-mail messages every day. When I was in college, no one had a cell phone, but Lulu was able to pull up SpongeBob videos on Netflix on my phone before she turned two. Technologies both constitute and disrupt our sense of the familiar. We grow accustomed to new things; they become normal parts of life. The industry is banking on us getting used to fracking.

Dostoevsky has a character say, "Man can get used to anything, the beast!" [17] I like that quote, because it suggests there is something less than human, something irrational, about constantly changing our schemes of familiarity. As we proceed headlong with innovation, we aren't just adjusting the means to some preestablished ends; we are adjusting the very goals themselves. What we call progress is like the Red Queen's race in evolution, the idea that an organism must constantly adapt just to survive in a changing environment. Sure, we'd be miserable now without television and cell phones, but that doesn't mean humans in the past were miserable because they were deprived of these things. When she was a kid, Gram Deal didn't sit around pining for the Internet. Our wants and needs grow right along with our new technologies.

As I talked more with Mark and others in the industry, I slowly came to see something nihilistic about the proactionary worldview. It has no goal; as Nietzsche says in "The Parable of the Madman," "Are we not plunging continually? Backward, sideward, forward, in all directions?" By contrast, I think there is something serenely Buddhist about the precautionary worldview: to keep the present order intact is not just a prudent stance in the face of uncertainty; it also liberates you from taking on the burden of more desire.

The term "proactionary" may be a recent coinage, but it is naming a worldview with deep cultural roots in the modern West. What is the proactionary archetype? Who is the anti-Frankenstein? Given the flaming faucets now associated with fracking, Prometheus, who stole fire from the gods, would make a poetic choice.[18] But Prometheus isn't quite right, because he is neither human nor divine but an intermediary of sorts. His story hinges on a discontinuity between mortals and the gods. He gives humans their distinctive identity *against* the gods' wishes. For the proactionary archetype we need someone who embodies the *continuity* between the human and the divine.

For this reason, Goethe's Faust is the best icon of the proactionary. It's not just that he makes a wager with the devil, symbolizing the core proactionary willingness to court danger for reward. And it's not just his restless striving for knowledge, which is so boundless that he says the devil can kill him should he ever find a moment in life that he enjoys so much that he wants to stay and savor it: for the proactionary, "enough" is a dirty word.

Rather, the most important thing about Faust is the blurring of lines between the human and the divine. This metaphysical slippage emerges in the Middle Ages (when the legend of Faust began) with the theologian Duns Scotus and his notion of *univocity*, the idea that humanity and God are only different in degree, not in kind. With this move, Scotus opened up the mental horizons of humanity. He cleared a space for utopian thinking and made possible the scientific revolution (Francis Bacon was a firm believer in univocity) with its hypothetical and *experimental* approach to nature. It is an approach dependent on a sense of humans as cocreators capable of bringing what is not-yet-here into being.[19]

I tried to read Goethe's *Faust* in German during college—a nearly impossible task, though I still remember some of the opening monologue: *Da steh ich nun, ich armer Tor!* The key moment for us, though, comes in scene 3, when he reads the Gospel of John in the New Testament. After Mephistopheles makes his bet with God (that he can tempt Faust away from the path of righteousness), we see Faust in his study crammed with books and instruments. His frustrated desire to know everything has driven him to the brink of madness and suicide. His quest for knowledge is not cold and scientific; it is an impassioned hunger, an erotic lust.

Mephistopheles visits him disguised as a howling, hyperactive dog, which only further grates on Faust's nerves. Jittering around the room, he suddenly opens the Bible to the line "In the Beginning was the Word." The "word" here comes from the Greek *logos*, and the precautionary mindset would translate it as "In the Beginning was Knowledge." But watch how Faust gives a proactionary reading of this phrase:

It's written here: "In the Beginning was the Word!"
Here I stick already! Who can help me? It's absurd,
Impossible, for me to rate the word so highly
I must try to say it differently
If I'm truly inspired by the Spirit. I find
I've written here: "In the Beginning was the Mind."
Let me consider that first sentence,
So my pen won't run on in advance!
Is it Mind that works and creates what's ours?
It should say: "In the beginning was the Power!"
Yet even while I write the words down,
I'm warned: I'm no closer with these I've found.

The Spirit helps me! I have it now, intact.
And firmly write: "In the Beginning was the Act!"[20]

In the beginning was the Act! Now, that's the proactionary spirit in its most concentrated form—even more pure than Kant's Enlightenment-defining dictum: "Dare to know!" For Faust, we don't passively receive the truth, and we don't try to figure out the truth before we jump into the business of living. No, we make things true through our actions, our experimental interventions in the world.

PRECAUTIONARIES ARE WORRIED about the unintended consequences that can result when we act before having a full understanding. But here's how a proactionary would respond: Of course we act with incomplete knowledge. How else can we act!? Individuals and even big institutions like the state can never comprehend the complex whole of society.[21] We can't consciously or intentionally move toward a planned goal, because important things will always elude our limited minds. So we run a little experiment here with incomplete knowledge. It sends ripples out, pinging the complex network, letting other actors know how to respond from their own little corner of society. We learn, adapt, evolve.

Even in the case of medicine, the proactionary will say, we eventually market drugs with something far less than complete certainty. The dictum "First, do no harm" cannot mean "Do nothing." If we were required to know everything about every product we manufacture prior to commercializing it, we'd never make anything. Sure, that means we live with risk, but that's the price of progress. And we also risk a great deal by *not* acting.[22] If we wait for a perfectly safe drug, we'll die from the disease. If we wait for a perfectly safe sunscreen, we'll get cancer from sun damage. If we wait for a perfectly safe energy system, we'll be stuck in the Stone Age. We have to set aside most of the complexity of nature if we are ever going to master it for our benefit. Yes, we'll overlook something important, but when that happens, solving the resulting problems will become our new focus.[23]

Advocating something like a proactionary mindset, the philosopher William James once noted that the precautionary view "is like a general informing his soldiers that it is better to keep out of battle forever than to risk a single wound. Not so are victories either over enemies or over nature gained."[24] No pain, no gain.

When we translate this spirit into policy, it forms the basic paradigm of "innocent until proven guilty," where we sort out the unintended impacts of a technology *after* commercialization.[25] Through the Toxic Substances Control Act (TSCA), for example, the Environmental Protection Agency keeps tabs on over 80,000 chemicals. Manufacturers are rarely required to provide data about health impacts prior to putting their chemicals on the market.[26]

Principles like precaution and proaction are not recipes. They don't turn democracy into a cookbook.[27] In the actual guts of policymaking, there will always be the need for judgment. For example, how much knowledge do we need prior to acting? There's no formula for answering that question, which shows how proaction and precaution exist along a spectrum. After all, no one is espousing utter paralysis by precaution or absolute recklessness by proaction. Any decision will entail some mix of precaution (we should know at least *something* about a technology prior to commercializing it) and proaction (it's just that we can't know *everything* about it). What I'm arguing is that the judgments one makes—where one falls along the spectrum—are driven at least partly by underlying worldviews.

Socrates stopped everyone and forced them to think. You shall not proceed on unexamined assumptions, he said. That's pretty much the precautionary approach to innovation, and it's precisely what Dr.

Frankenstein was doing to that eager young ship captain brimming with Faustian desire: don't go down that path, you haven't thought it through, you'll regret it.

John Siegmund's proposal for technocracy has some things in common with Socrates's proposal in *The Republic* for rule by philosopher kings. Most importantly, neither regime is fond of free-market democracy, which, Socrates noted, basically amounts to rule by idiots (those who are ignorant). Of course, the difference is that engineers wouldn't question our consumerism. They'd just service it more efficiently. Philosopher kings, by contrast, would keep our animal urges in check so that we would have a fighting (albeit still slim) chance of becoming excellent human beings. John wants rule by experts-of-the-means. Socrates wants rule by experts-of-the-ends. Engineering kings make us more efficient idiots; philosopher kings make us less idiotic.

For Adam Smith, both ideals fail because they rely far too much on centralized planning: whether engineer or philosopher, the king would have to be all-knowing, which is impossible. The beauty of the invisible hand is that you only need to pursue your self-interest. The public good will automatically result: "He intends only his own gain, and he is in this, as in many other cases, led by an invisible hand to promote an end which was no part of his intention." [28]

That's the same faith behind the proactionary wager: from narrowly intended goals we will bring about the common good. The precautionary principle is an attempt to force science and technology toward some prearranged utopia. Max More thought that was ludicrous. So did Smith, who called it "folly and presumption." [29] We can't steer open-ended processes of discovery and innovation toward some tightly controlled and predicted outcomes. [30] Like those methane molecules lured upward out of the rocks, we just have to go with the flow and have faith we will work things out along the way. It's the Zen of proaction.

Smith was the Faust of philosophers. He argued that it may look like humans want convenience and comfort, but what we actually want is to be perpetually busy perfecting the machines that would make us comfortable should we ever stop to enjoy them.[31] The human essence is restlessness. We'll never say "enough."

Smith did much to wed the proactionary spirit to the economic system of capitalism. In so doing, he was following the lead of John Locke. Locke thought that God gave the Earth to humans so that we could "increase" and improve our condition.[32] Before it is mixed with human labor, nature is mere "waste." Nature offers potential abundance, but it remains constant within set limits and cannot, of its own power, *increase*. It's kind of spooky how Locke's conception of nature as waste is reflected in the primary mission of the Texas Railroad Commission, to "prevent waste." They think of "waste" as leaving hydrocarbons in the ground.

Transforming waste (er, nature) into wealth requires human ingenuity. But not everyone is equally smart. So as we parcel nature into segments of privately owned property, some will fare better than others. We'll see the rise of economic inequalities. Those like Mitchell and other entrepreneurs who are more naturally gifted and industrious will acquire more wealth than others. This inequality is justifiable, Locke argues, because these men will have created new conditions of plenty for everyone. When they cultivate ten acres of farmland (or produce ten acres of oil and gas minerals), they increase tenfold the value of an otherwise miserly nature. In essence, they give ninety acres to humanity. They have created value. Sure, they live opulent lives among the "1 percent." But, as Locke put it, a king in America (still stuck in the state of nature during his days) is worse off than even a petty day laborer in an industrial civilization like England. Most folks in Denton may not be wealthy (the average per capita income is about $25,000), but they do lead comfortable lives made possible by the energy that people like Mitchell turned from inaccessible waste into actual heat, light, and video games.

Locke notes that to ensure *everyone* is better off, we must especially protect the possessions of the "industrious and rational" few from the "fancy or covetousness of the quarrelsome and contentious." He believes that the wealthy industrious class is so important for progress that he calls that prince "godlike" who protects their property rights to the fullest extent.[33] It's enough to warm the cockles of Ayn Rand's heart. What Locke may not have anticipated is just how pronounced inequalities would become in capitalist systems that tend toward monopolies.[34] And he surely could not have foreseen the immense scale and impact of industrial technology.

With Smith and Locke, and their central concepts of the invisible hand, private property, and justifiable inequalities, we can no longer avoid the way the proactionary embrace of innovation tends to blend naturally into corporate capitalism. Smith and Locke argue that if we are to make the Faustian wager of transforming the world, some will win and some will lose. They, like the libertarian economist Julian Simon, claim that everyone will be made better off in the long run as standards of living rise. That may be true, but there is no denying that many are harmed along the way—those guinea pigs of progress. And it remains an open question just how long we can sustain our now planet-wide technological gamble.

Another way to put the point is to argue that proaction, an admirable ideal on its face, is easily corrupted by an economic system predicated on massive inequalities of wealth and status. In such a system, those who are harmed in real-world experiments are most likely to be disenfranchised and silenced. This is the conundrum: the proactionary spirit has given rise to an economic and political system that undermines what was most noble about the ideal of progress through freedom. We might say the spirit has been adulterated—a noble quest for transcendence and self-expression has run aground in the mire of greed and consumerism. To put it crudely: Faust was out for love, while the frackers are out to make a buck.

Max More notes that the key to the proactionary principle is objectivity in the assessment of risks. Yet (especially American) neoliberal capitalism short-circuits any principled appraisal of risks. Those who profit from the thousands of technologies that are allowed to bloom will inevitably distort the risk assessment process. The longer the real-world experiment runs, the more money and power they will have to ensure that no weeds are ever found. Corporations increasingly control real-world experiments—their political oversight, their media coverage, and the production of scientific knowledge about them.[35] They are adept at creating and magnifying uncertainty to delay renovations.[36] They even control the museum exhibits, which abandon any attempt to critically assess fracking in favor of cheerleading. In short, it is naïve of More to think we can create billions of dollars of vested interests around a new technology and then conduct an "objective" assessment of it.

The final scene of Goethe's play witnesses Faust's forgiveness. Despite all the suffering and anguish he unleashed on Earth, he ascends to heaven in the presence of God who takes the form of the eternal feminine. The angels declare:

> *He who strives on and lives to strive*
> *Can earn redemption still.*[37]

It's supposed to be an uplifting message, but in the capitalist context of fracking and other disruptive innovations, a cynical reading is all too easy. Only the rich and powerful will be able to escape the mess their frenetic experimentation has made of the planet. Redemption is reserved for those who can afford it.

ON JANUARY 6, 2014, the members of DAG convened at Rhonda's house for what would turn out to be our most important meeting. Hydraulic fracturing was going to begin in the Meadows neighborhood the next day. The previous night, some volunteers had organized a candlelight vigil at City Hall to commemorate that mournful occasion.

We sat in our usual spots. And, as usual, Rhonda had prepared too much food, set out two pots of coffee, and stocked her beer fridge. There was a good reason we met at her house! We had learned not to meet at my place. The very few times I played host, people had to clear away enough toys just to find a seat. One time, Sharon lost her water bottle in the mess at our house and it went AWOL for months.

A nervous tension filled the air. It was time to answer the ultimate question: To ban or not to ban? We had been struggling with the exact wording of a possible ban for a long time. A leading approach then was the community rights framework, which asserts broad, fundamental rights to clean air and water as a justification for banning fracking.[38] Here's a brief example:

> Right to Clean Air. All residents, natural communities and ecosystems in The City of Broadview Heights possess a fundamental and inalienable right to breathe air untainted by toxins, carcinogens, particulates and other substances known to cause harm to health.[39]

Though we respected this framework and those who were using it, Cathy, Sharon, and I didn't think this kind of approach made sense for us. If a city really took all those rights claims seriously, then it would have to ban just about everything, including cars and lawn fertilizers. The sweeping rights claims were an indictment not just of fracking but of the modern world.[40] We also couldn't couch a ban as a zoning ordinance (as was the case, for example, with the ban successfully defended by the town of Dryden, New York), because zoning issues, by Texas law, cannot be settled via citizens' initiatives.

A lawyer (who shall remain anonymous, but to whom I am grateful) helped us draft some language to fit our situation. It simply treated hydraulic fracturing as a business practice that was incompatible with the health, safety, and welfare of Denton residents.[41] Now we had something concrete to vote on.

Cathy made a motion for DAG to approve the draft language, turning it into a citizens' petition to ban hydraulic fracturing in Denton. Ed quickly seconded the motion. Everyone then looked over at me, knowing I was the one who had been holding on to reservations the longest. I just stared at them for a moment from my reclining chair. It was a difficult, uncomfortable spot for a philosopher to be in. I could have hemmed and hawed about it for much longer, but I saw the conviction in the faces of Cathy, Carol, Rhonda, and Ed. It was time to draw a line in the sand.

"Let's do it," I said. It was something of a leap of faith, because it's so hard to know what's right. What I did know was that fracking in Denton had become an abject policy failure. That was certainly the case when evaluated by precautionary standards, but it was even a failure according to proactionary terms. The problem wasn't that fracking in Denton was a real-world experiment; it was that fracking was an *unjust* real-world experiment.

The motion passed unanimously. Though it wasn't clear to us at the time, this meant that if we were going to get our petition into the city code as an ordinance, we had ten months to turn our five votes for the ban into roughly 13,000.

I remember coming home that night to tell Amber the big news about the ban. The kids had (as usual) made a mess of the house that day, so we ended up talking about all the things a ban would entail—lawsuits, attacks from the industry, fund-raising, mobilizing voters, the implications for my tenure at UNT—while we mopped up SpaghettiOs and put away crayons.

The next day, I texted Shane Davis, whom I had met earlier in Fort Collins, Colorado. Widely known as the original "fractivist," he led much of the resistance to fracking, including municipal bans and moratoriums, in Colorado. He's the one who taught me about the importance of language in the fracking debate, as he instructed me to always refer to condensate tanks as "toxic waste tanks." The philosopher Ivan Illich called neutral-sounding terms like "condensate tanks" the "minting stocks of public intercourse."[42] They configure a world of control and calculation. To use them is to unconsciously inhabit a world that is prearranged for you by experts and those in power. It puts you on the defensive and on their turf right from the get-go.

In addition to a sharp mind, Shane has what the ancient Greeks called *thumos*, the kind of spiritedness you find in warriors or wild horses. When he got my text about the ban, he wrote back a congratulatory note and included a line from Sun Tzu's *Art of War*: "Attack is the secret of defense; defense is the planning of an attack." I had hoped the process of seeking a ban would enrich the democratic conversation, but I would come to realize how naïve I was and how right Shane was. Rather than a conversation, it would be a war. In addition to *logos*—the arguments for the ban—I was going to have to work against the grain of my mild nature and cultivate some *thumos* of my own.

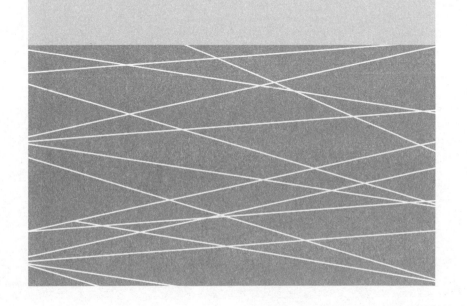

Part II

GUINEA PIGS
OF THE SHALE

PLAYING NICE IN THE SANDBOX

I T WAS SIX AGONIZING WEEKS BETWEEN THE TIME
DAG decided on the fracking ban (in early January 2014) and the
time we made the public announcement. The ban was a secret between
Texas Sharon, the DAG Board Members (Ed and Carol as treasurers,
Rhonda as secretary, Cathy as president, and me as vice president),
and our newly formed social media team (Angie Holliday, Nikki
Chochrek, Tara Linn Hunter, and Christina Bovinette). During that
time, we were starting to plan for the campaign, from gathering signatures on the petition to registering voters and mobilizing a get-out-the-
vote strategy. We also had to form a political action committee, which
we called Pass the Ban, and we needed to establish the terms of our
relationship with Earthworks via Sharon.

We were also struggling with how to brand the campaign. Initially,
we thought of "Don't Frack with Denton," but that sounded too negative. As I recall, it was Tara Linn who came up with "Frack Free Denton," which we all loved right away. Her husband, Matthew, who is
an artist and musician like Tara Linn, put together our logo, which

we put on our website, T-shirts, and Facebook page, and which would eventually adorn all of our advertisements and yard signs.

Coincidentally, it was also in early January when five hundred people packed an auditorium in the nearby town of Azle (rhymes with basil). They had come hoping that state oil and gas regulators would explain to them why there had been over thirty earthquakes there in the past two months. The regulators said they didn't know what was causing the unusual seismic activity and promised to "get some state research done." [1] Most people in the frustrated crowd were convinced it was caused by fracking, especially injection wells that store fracking wastewater. Shortly after that meeting, Texas Sharon joined many of Azle's citizens on a "frack-quake bus ride to Austin" to tell their stories to state legislators. One woman told Sharon that there was enough subsidence at her home that you could see unpainted portions of a fence that used to be underground. It was as if the earth were dropping away from below and things were left rootless, floating.

As the Azle earthquakes made headlines in January, DAG hosted an educational event at the Denia Recreation Center in Denton. In our presentations, we shared the growing scientific literature about the harms of fracking and showed how fracking would continue to occur all over Denton neighborhoods right next to homes despite our current city ordinances. Afterward, the little gym packed full of over a hundred people started buzzing with questions. A military officer who had just moved to town asked: "What are we going to do about all this?!" We knew exactly what we were going to do, but all we could say was "Stay tuned." We were not ready yet to unveil the strategy of seeking a ban.

Outside of our inner group, the only other soul who knew about the ban was the *Denton Record-Chronicle* reporter Peggy Heinkel-Wolfe, who had been covering urban fracking longer than anyone else.[2] Peggy had earned the scoop on this story. Her article announcing the petition drive to gather signatures and put the ban on the ballot ran in the *DRC* on February 18, 2014.[3] Two days later, a Thursday, we held our petition launch party at Sweetwater Grill and Tavern just a block south of the courthouse square. We came early to set up the back room, wearing our new black-and-white Frack Free Denton T-shirts and full of nervous energy. How many people would come? Would some arrive just to heckle us? Cathy couldn't stop pacing back and forth. Sharon was relentlessly biting her fingernails. Ed and Carol were flittering around just trying to keep themselves busy. Even the unflappable Rhonda ordered a beer to calm her nerves.

The first people to arrive were about a dozen volunteer petition circulators. Their job would be to gather signatures from registered voters in the city. They would go door to door in Denton's neighborhoods and to the community market, music events, and local restaurants and bars. We gave them each a clipboard with a long sheet of paper that had lines for signatures on the front and the full wording of the proposed ordinance banning hydraulic fracturing on the back. Cathy led the training session, instructing the volunteers on the rules of a citizens' initiative. Little did she know then how much sleep she would lose worrying about those petitions, which, once they were all signed and notarized, she would keep locked tight in a fire safe by her bed until the day came to turn them in to City Hall.

As the official time of the launch party arrived, people started pouring in, and soon the room spilled over with more than two hundred enthusiastic supporters. The sight of it nearly brought Cathy to tears. I'll admit I got a little emotional too. It had been a long time banging our heads against administrative walls, hosting meetings, and handing out flyers. There was something uplifting about that crowd and the bright lights turning on atop news cameras from Dallas and Fort Worth. We had tried to protect neighborhoods according to the terms of the bureaucracy. That hadn't worked. Now, with the petition, *we* were setting the terms. For the first time in a long while, this felt like real democracy.

Ed was always the one to call our meetings to order because he has a booming voice acquired from years spent teaching drummers. As he said a few words of introduction, I took a seat at the front of the room facing the huge crowd of people standing shoulder to shoulder all the way to the back. Ed turned the floor over to Cathy, who suddenly looked calm and completely in her element. She said we had just become a target of the oil and gas industry. "They're going to try to intimidate us," she told the crowd, "but you tell me, do I look scared?" She said it with a steely, cold look in her eyes that I had never seen before. Everyone hollered and clapped. In that moment, Cathy matured into our fearless leader.

Calvin Tillman, the former mayor of nearby Dish, who had been an ally of ours from the beginning, spoke next. He stood with his son Josh and told the crowd that the ban wasn't about partisan politics; it was about health and neighborhood integrity. A kind of hush fell over the crowd to match his soft-spoken demeanor: "This is about your right to feel safe in your own home. I ask you . . . what's that worth to you?"

The next speaker was Maile Bush from the Meadows neighborhood, where her home was sandwiched between three gas wells that had just been fracked by EagleRidge Energy. Maile (rhymes with wily) has that mixture of dogged determination, intelligence, and fierce mama bear attitude that can turn an average citizen into a crusader. Over the past few months, she had devoted her precious spare time to learning about fracking. She peppered Cathy, Sharon, and me with questions over e-mail and during our neighborhood meetings at Fire Station #7.

"My name is Maile, and I'm not a terrorist; I'm a mom," she told the crowd at the launch party. By this point, the audience had grown more raucous and even larger, now wrapping around behind her so that she was surrounded. I had to stand on my chair to see, and even then I could just catch the occasional glimpse of her curly brown hair and the side of her glasses as she said, "I didn't set out to be a fracking activist; I'm just trying to protect my kids."

Maile told them how for months during 24/7 drilling and then hydraulic fracturing, she and her husband, Jason, had to keep their young children, Kaden and Cassidy, indoors. Even then, they still developed headaches, coughing, and nosebleeds. EagleRidge put a compressor on the southern frack site in her neighborhood, which was noisy and emitted noxious odors. Kaden continued to get at least one nosebleed every week (a symptom he had never had before). It got to the point where he was hiding them from Maile, because he didn't want to upset his mom. And all this is not to mention the constant noise and light that often kept them up at night. It was a common story in that neighborhood, with other young mothers describing similar symptoms in their homes and saying that it was like "living in hell" and that "we felt like prisoners in our own home."[4]

Maile's crack about not being a terrorist was prompted by a recent incident involving Mark Grawe (the head of EagleRidge Energy). He had walked into a meeting in the nearby town of Mansfield, where his wells were also causing controversy, with armed security guards and claimed that several folks in Denton were on the US Department of Homeland Security's watch list of suspected terrorists.[5]

That would have seemed laughable to me if it hadn't been for my visit from the FBI almost exactly two years prior to the petition launch party, back when the task force was deliberating about the city's drilling and production ordinance. On February 8, 2012, I had just walked back to my office from teaching at another building on campus. As I turned the corner, I saw two men standing outside my door. When they told me they were with the FBI and the Dallas Police Department, I got that sinking feeling that happens when you see the flashing lights in your rearview mirror.

We wound up having the most bizarre thirty-minute conversation in my office, the walls of which were plastered in MG's artwork (he had turned four just two days earlier). The FBI agent asked:

"Have you heard anyone mention IEDs in relation to fracking?"

"What?" I stammered.

"You know: improvised explosive devices."

"Like the kind they use in Iraq?" I asked in disbelief. Did they really think someone was going to start bombing gas wells?

I said, "No one is talking about bombs. Most people at DAG events are not even comfortable holding signs at demonstrations. Some of them are actually in favor of drilling . . . but, you know, not right by their homes. They just feel disenfranchised."

Then the Dallas PD officer said:

"I see you cover eco-terrorism in your class."

"Um, yeah," I replied, now really starting to sweat. "We actually just read the *Field Guide to Monkeywrenching* that advocates for the sabotage of wilderness-destroying equipment."

I explained that I like to use essays on civil disobedience, like Martin Luther King Jr.'s "Letter from a Birmingham Jail," to begin my ethics classes. They cut to the core questions: What is a just law, and how can you tell it from an unjust law? What should you do when your conscience tells you a law is unjust?

"Hmm. Yes. Well, we're concerned that a Seattle-style radicalism might migrate south, following the fracking controversy," the Dallas police officer said matter-of-factly.

"Uh-huh. Really!? . . . Hmm . . . ," I spluttered. Did they really think I was radicalizing students?

"We just want to make sure that everyone is playing nice in the sandbox," the FBI agent said with a smile.

"Oh," I replied, beginning to wonder if I had said enough.

That phrase about the sandbox was repeated at least five times. In fact, the whole thing was like running laps at a track; we kept covering the same ground over and over. I was later told this is a common interrogation technique; they wanted to see if my story would change.

Next they asked about one of my former students at UNT, Ben

Kessler, an ex-marine and member of the environmental activist group Rising Tide. Ben had been arrested about a year earlier at a protest in Washington, DC.

"What's your take on Mr. Kessler?"

"He's a great philosopher," I said.

They just stared at me. So I added, "I can't imagine he'd ever blow something up. Ben is deeply committed to nonviolence."

The FBI agent handed me his card, saying, "We'd like you to keep your ear to the ground and let us know if you hear of anyone threatening violence. If so, we want to take their temperature."

That was the other oft-repeated phrase: "take their temperature." I imagined them snooping around with thermometers, stalking activists.

At this point I couldn't resist saying, "I do tell my students to break the laws . . . but only the unjust ones." I was thinking about how moral heroes like Oskar Schindler would be the bad guys if we're supposed to follow the rules no matter what.

They just gave me a bemused look. And the FBI agent said, "I sympathize with the activists and affirm their First Amendment right to protest. It's just that . . ."

". . . we want to make sure everyone is playing nice in the sandbox," the Dallas police officer finished the sentence for him.

After they left, I wondered if my temperature had just been taken—and what the reading was.[6]

It was enough to make me paranoid, or at least less quick to roll my eyes at Texas Sharon when she tells me her cell phone has been tapped (and this was before Edward Snowden exposed the extent of the government's surveillance capacities and interests). In fact, it was Sharon who first showed how not just the government but oil and gas corporations think of local resistance to fracking in terms of terrorism. A few months before my visit from the FBI, she caught an industry representative from Range Resources (the company that fracked by the purple park) on tape boasting about the "former psy-ops folks" that work for him.[7] Psychological operations, or psy-ops, are tactics for destabilizing a local population in the face of an invading army, or in this case, invading corporations. Another industry representative spoke of local resistance to shale gas and declared, "We are dealing with an insurgency." He advised industry representatives to obtain the US Army/Marine Corps Counterinsurgency Manual and Rumsfeld's Rules to learn how to combat local citizen groups opposed to fracking.

Maile's story at the petition launch party made me think about my visit from the FBI again. The fracking that happened in her neighborhood was the result of years spent advocating for changes from within the existing legal framework that systematically favored the industry. After Maile spoke, Rhonda handed Cathy a pen to put the first signature on the petition as cameras flashed all around us. With Cathy's signature, we had officially stopped playing nice in that sandbox.

As Cathy handed the pen to me, Peggy was standing on a chair taking a picture for the *DRC* with her phone and tweeting @adambriggle and @frackfreedenton: "Best thing about being a journo is that front row seat to history."

IN REAL-WORLD EXPERIMENTS, like the one that happened in Maile's neighborhood, some people are going to be exposed to (often unknown) risks of harm. Much the same can be said of scientific field experiments that involve human subjects. Think, for example, of clinical trials that test new drugs or medical devices. In the case of both fracking and clinical trials, the expected gains can only come about by subjecting people to potential harms. In the former case, it is those, like Maile's family, who live, work, or play near the industrial sites. In the latter case, it is the research subjects who take the experimental drug.

In the case of clinical trials, the experts (or the government) could determine when an experiment is safe enough or important enough and, at that point, compel people to participate. In fact, we used to do that. For example, from 1932 to 1972, scientists and physicians working for the US Public Health Service withheld antibiotic treatment from poor African American men in Alabama suffering from syphilis. It was an experiment to see how the disease progressed over time. In this now infamous "Tuskegee syphilis experiment," the men were kept in the dark about the purpose of the experiment, the disease they had, and the existence of a cure.[8]

Since then, the United States has made a very conscious decision: scientists do not get to force people to risk bodily harm in the name of social progress. Indeed, this decision has been made around the world in various declarations, laws, and codes of conduct. For the most part, modern society has rejected the premise behind the Nazi experiments on concentration camp prisoners, namely, that some people are mere means to be used for social ends. Even if it would serve the greater good, no one gets to treat you like a guinea pig.

That is, unless you give your consent. The first line of the 1947 Nuremberg Code, an international moral rejection of the Nazi experiments, reads, "The voluntary consent of the human subject is absolutely essential." The later US Belmont Report (published in the *Federal Register* in 1979), a reaction to the Tuskegee experiment, similarly ranks informed consent as the top priority for judging the moral acceptability of experiments involving human participants.[9] This is rooted in a core Western principle of respect for persons, which is in turn based on the thoughts of Immanuel Kant.

We should add Kant to Locke and Smith as one of the forefathers of proactionary thought. He believed that humanity, at least its mature form, is defined as a kind of daring. According to Kant, we have the moxie to transgress boundaries laid out by the so-called natural order of things and by stale traditions. It's this kind of daring that is now widely celebrated in our entrepreneurial class, including those "frackers" who pioneered the shale revolution.[10] But for Kant there is one limitation to the bold pursuit of knowledge: it must never violate the freedom and autonomy of fellow human beings. Kant would never let his morality be corrupted by a cold utilitarian sacrifice of the few for the many.

These deep roots are why I think a proactionary ethic for innovation and engineering must mirror the requirement for informed consent found in medical research.[11] *Those who are most vulnerable to the harms posed by fracking should have the greatest say in decisions about whether and under what conditions fracking should occur.* That's my first criterion for ethical innovation.

And judging by it, the fracking in Maile's neighborhood was an ethical failure. Although the gas well sites predated the construction of the homes, no one was adequately informed. R. L. Adkins Corporation originally drilled two vertical wells (one on the north site and one on the south) in 2003. Back then, the Meadows at Hickory Creek neighborhood didn't exist. Ten years later, after Maile and her neighbors were living there, EagleRidge got permits from the Texas Railroad Commission to convert the two existing vertical wells into horizontal ones and to add two more wells to the north site. But due to the vested rights laws enshrined in chapter 245 of the Texas Local Government Code, all this "new" activity was not considered new at all. Rather, it was just part of an ongoing project first started with the permits obtained by R. L. Adkins in 2003.

That meant that EagleRidge's plans did not require a public hearing, which is an important vehicle for informed consent when it comes to fracking (or other land use matters). It also meant that EagleRidge was not required to get a new specific use permit (SUP) for their project. By stipulating criteria necessary to make a nonconforming use compatible in another zoning category, an SUP is what allows the industrial land use of fracking to occur in Denton neighborhoods (which are obviously zoned residential, not industrial). The 2003 SUP stated that the gas wells were "compatible with and not injurious to the use and enjoyment of other property . . . within the immediate vicinity."[12] But that's because there was no one living on property in the immediate vicinity back then! The people, like Maile, who had since moved to that area got no say in the decision to frack there in 2013, because that decision had been made for them ten years ago by city officials under vastly different circumstances.

Another mechanism of informed consent in such situations is disclosure for new homebuyers. This too was almost entirely absent. When they are not being actively fracked, gas wells can be fairly easily overlooked. So several people didn't even know they were there. Two residents in that neighborhood told me that the sales agent for the developer, D. R. Horton, lied to them. When they asked what those "things" were, the sales agents said they were just water wells or they were abandoned gas wells never to be developed again. The only indication that fracking might happen at those sites in the future was a small paragraph tucked into the hundreds of pages signed at the closing of the home sales. Not surprisingly, no one read it. Yet the city's ordinance was set up under the assumption that people were making fully informed decisions when they moved to neighborhoods with preexisting gas wells. That's the reason the "reverse setback distance" (between old wells and new homes) is just 250 feet (with caveats that allow that to be further reduced to less than 200 feet).[13]

We don't bury the risks of an experimental drug in mountains of paperwork to be signed only after one has agreed to participate in a clinical trial. Rather, a researcher who is knowledgeable about the drug explicitly discusses the risks before any decisions are made. Something like this should have happened in Maile's neighborhood too. Indeed, even as we went forward with the proposed ban, City Council tried to find ways to improve disclosure—perhaps through a brochure analogous to the one given to potential homebuyers when there is lead paint present in the house. The exact mechanism was murky, but the general idea was crucial, because everything would have been different had the people in the Meadows knowingly and freely consented to the frack-ing. It wasn't just that they were exposed to the risks; it was that they didn't choose to bear the risks.

When we announced the ban, I knew disclosure was a key piece of the puzzle, but I couldn't see any way to make it work. Even if we could solve issues of enforcement, there was a more basic problem: How could someone give "informed" consent when there was so lit-tle information to go on? The first major compendium of scientific studies on fracking came out a few months after our petition launch party.[14] Noting the "fundamental data gaps" still remaining, it called for a moratorium on shale gas development. Fracking is like a drug that skipped over all the clinical trial stages—in vitro testing, ani-mal testing, small human trials—to go directly to mass commercial-ization. In other words, the conditions are not in place for people to give their *informed* consent. How could they, when the industry uses secret chemicals, isn't required to inventory their emissions, and often includes nondisclosure clauses in their leases that prevent people from learning from one another's experiences?

THE SIGNATURE-GATHERING PHASE of our campaign (February to May 2014) coincided with City Council election season. Mayor Mark Burroughs, who had participated in a DAG-sponsored candidate debate two years earlier, was term-limited out. Chris Watts, whom I had interviewed about fracking near the purple park, was running for mayor against Jean Schaake, a fellow faculty member and dean at UNT. Before we decided on the ban, I had gotten to know Chris better and joined his steering committee.

In April, well into the Frack Free Denton campaign, I wrote a blog post endorsing Dalton Gregory in his reelection bid for City Council. But I didn't run my decision by the DAG board members, many of whom thought Dalton was the wrong choice, because they saw him as weak on the industry. Actually, to be more candid, they despised Dalton.

Cathy called me on the phone, furious about my endorsement. Sharon gave me that laser death stare that I hoped never to have directed toward me. Ed and Carol were angry and disappointed. Maile and some others in the Meadows neighborhood took it as a personal slap in the face, because they thought Dalton was part of the reason fracking had happened so close to their homes. For several weeks, they all gave me the silent treatment.

This happened on Easter Sunday, and I'll never forget that day and how horrible I felt as I got off the phone with Cathy and watched the dark clouds spill rain on our back patio. DAG almost fell apart. I almost derailed the whole campaign.

And it was all because I didn't get the consent of the group. That episode was incredibly formative for me—it made me realize something that had become so close, so ingrained, that I couldn't see it until then: I was an autonomous academic, free to speak my mind however and whenever I wanted. But I had to set that identity to the side. At least until the vote on the fracking ban, I was not a lone wolf. I was part of a political PAC, and we had to move in lockstep if we were going to win. For as important as Kant's idea of autonomy is, it is not the only value in our lives. There's something to be said for solidarity.

For a while, I grumbled and complained to Amber about this situation—it felt like censorship. But once the industry attacks started coming, I could see the necessity of our unified front, and I eventually came to revel in the camaraderie of the group. It felt like the old days being a member of a hockey team.

I never retracted my endorsement of Dalton (who ended up winning by a landslide)—I never doubted that he was the right choice—but I did take down my blog post. For a couple of weeks, I was in exile from the group as we worked through hurt feelings. But this gave me a chance to step back and reflect more on the ethics of real-world experiments. I thought about how fracking more generally—not just by the purple park or in Maile's neighborhood—fails the first criterion of informed consent. The sandbox, or legal framework, is set up in such a way as to systematically disenfranchise many of those who are vulnerable to harms. As one of Maile's neighbors said, "You kind of just put up with it because you don't feel like you have a say." [15]

To help me make this point, let me divide the sandbox into two parts and call them the "private" and "public" dimensions of fracking policies. The private decisions pertain mostly to the leasing of mineral rights. The public decisions pertain to federal, state, and local regulations.

First, consider the private decisions. Those who own the minerals hold the lion's share of power to dictate whether or not to frack. One way to achieve the condition of informed consent, then, would be to ensure that those living on the surface and exposed to the risks of harm own the mineral rights so that they are empowered to decide whether or not to accept those risks.

Yet in states like Texas that sever the mineral and surface estates, this often doesn't happen. The legal history of Texas oil and gas stretches back to hunting on nineteenth-century English manors. [16] Foxes, ducks, and other wild animals (*ferae naturae*) freely roamed across invisible and artificial human property boundaries. Drawing from the primordial tradition of possession as the origin of property, English courts decided that these "fugitive resources" should be considered the property of whoever is in actual possession of them at the time. In other words, if you shoot it on your land it is yours.

Lacking almost any scientific understanding of mineral resources, late nineteenth-century American state courts similarly treated oil, natural gas, and groundwater as *ferae naturae* and analogously applied the rule of capture to demarcate competing ownership interests. In other words, if you drill it on your land it is yours. In 1904, just three years after the historic Spindletop oil gusher in southeast Texas, the

Texas Supreme Court lamented that it was hopeless to try to develop any more precise rules for a natural phenomenon so "secret and occult."

This meant that as the oil and gas boom took off in Texas, it was a total free-for-all. In 1935, a landowner sued his neighbor because his neighbor's well was draining the mineral resources under his land. But the Galveston Court of Civil Appeals rejected his claim as meritless. After all, his neighbor was not stopping him from drilling his own well—both could tap into the same oil reservoir. If you wanted the minerals under your land, you'd better be the first to grab them. And if you saw your neighbor putting up a rig, you'd better do the same quickly. This, however, led to overproduction and waste, the prevention of which was the original justification for the Railroad Commission's jurisdiction over oil and gas.

The rule of capture created the legal framework by which the mineral and surface estates could be severed. Though it had been the working law of the land for years, in 1971, the Texas Supreme Court explicitly recognized an oil or gas interest as the "dominant estate." [17] It's dominant because it would be rendered wholly worthless if its owner were unable to access the surface above in order to exploit the mineral reserves below. The mineral estate only has one use. On the surface, if you can't build a Whataburger you can build condos or a home. But if you can't extract your minerals, then you can't do anything with your subsurface property.

At the Range wells near the purple park, there was only one Denton resident who owned any of the pooled mineral rights. He owned less than one half of 1 percent of the appraised value of the minerals and lived over three miles away from the actual site. The Rayzor family (whose members scattered across the country after the ranching days were over) owned the biggest share of minerals. Someone living in Arizona had the second-largest share. No one in the neighborhood across the street owned mineral rights. Even if they had, their minerals would not have been pooled in the lease, because the horizontal legs of these wells stretched away from their homes, to the northwest. On the surface, though, the winds would bring emissions from that site directly over their homes.

No one in the Meadows neighborhood owned any mineral rights, either. Those rights were held by two businesses and two family trusts. This was about the norm for Denton. Working with Matthew Fry, my colleague at UNT, I had discovered that about $1 billion worth of appraised mineral values (a proxy for royalties) had been dug out of Denton's soils in the past ten years. About 70 percent of that was taken by the companies that invested the capital to make the production of value possible, like Devon, XTO, EnerVest, and EagleRidge. Only about 2 percent of the wealth belonged to the City of Denton. Another 1 percent belonged to actual Denton families (and most of this was concentrated in the hands of only about twenty families, including a couple of families that would soon form the PAC Denton Taxpayers for a Strong Economy officially opposing the ban). Absentee mineral owners held about ten times as much wealth as local residents.[18]

The largest single owner of Denton minerals is Robson Denton Development LP, based in Arizona, which claimed over $57 million in appraised value from 2003 to 2013, three times the amount owned by all local residents combined. Brian Baldinger, a former Dallas Cowboys football player who now lives in New Jersey, was the eighth-largest mineral holder, with over $4 million in mineral wealth. In other words, decision-making authority about whether to frack is dictated by the arbitrary history of property exchange rather than the ethically significant variable of exposure to harm.

It made me wonder how I would feel if the people in Krotz Springs, Louisiana, proposed a similar ban in the area where my family owned the minerals produced by Don Deal's wells. During the Frack Free Denton campaign, one absentee owner of Denton minerals sent an anonymous note to Sharon expressing how distraught she was about profiting from an industry that exposes children to toxic chemicals

near their homes. She said that she channels her royalties into efforts
to stop fracking. For her, it boils down to a simple question: How
much money would you take in exchange for risking the health of
children?[19]

Even the mineral owner's consent is not always necessary for a project to proceed. Thirty-nine states have some form of "compulsory integration" or "forced pooling" law that compels holdout mineral owners
into joining lease agreements.[20] The details vary by state. In New York,
for example, after 60 percent of the pool has been leased, the operator
can collect all the gas, even under the land of those who did not give
their consent.[21] In Texas, operators have invoked Rule 37 thousands of
times, a move that gives them the authority to drill against the will of
holdout mineral owners. In some cases, they don't even have to pay all
the mineral owners.[22]

The rationale behind such rules is to prevent the waste that would
result from drilling more wells on a patchwork of leased and non-
leased land. But it amounts to a kind of tyranny of the majority and in
extreme cases verges on theft. Even some staunch fracking advocates
like former governor of Pennsylvania Tom Corbett object to forced
pooling as an illegitimate version of "private eminent domain."[23]

There are other thorny issues. For example, what counts as consent in situations where someone is in dire economic straits? In some
depressed rural areas of Pennsylvania or North Dakota, people may
"consent" to fracking even though they might not have done so if they
could afford to say no. This is the same fuzzy border between consent
and exploitation that occurs with questions of payments to entice people to participate in drug trials or even to sell their organs.

Another major problem with fracking has to do with what it means
to be informed. Especially in the early part of the boom, many mineral
owners were signing leases with very little understanding of what they
were getting themselves into. Landmen are able to talk up the huge
amounts of money that could be made while downplaying all the fine

print in the leases, even though that fine print may cut into royalties or amount to a near total takeover of one's property.[24]

Frank Romeo, our old farmer friend from the days Amber and I volunteered on the farm in Pennsylvania, sent me a letter after he heard about Frack Free Denton on the radio. Frank was friends with many of his Amish neighbors, and he wrote that some industry landmen had been lying to them, taking advantage of their religious objection to initiating lawsuits. Some Amish farmers lost out on thousands of dollars in bad deals—money they would later need to mitigate the impacts of fracking on their land. But they didn't sue, because Jesus said to turn the other cheek.[25]

THOSE ARE THE "private" decisions on the market, but there are also public or regulatory decisions. Fracking is primarily regulated at the state level. As the forced pooling laws suggest, state regulatory agencies like the Railroad Commission are mostly concerned about promoting the development of mineral resources. They are the heirs of Locke's view that the function of government is to facilitate the technological exploitation of nature.

The Railroad Commission has a website with an interactive map that gives a telling insight into its subterranean perspective.[26] The gas wells at the purple park, in Maile's neighborhood, and at thousands of other sites on the Barnett appear as pentagons with long lines, representing the horizontals cutting through the shale 7,000 feet underground (an example of such a map can be found in the illustration insert). The surface, in other words, isn't just abstracted but actually made invisible. If you live in a community atop a mineral resource, the state doesn't see you. At least, it doesn't see you as a person in a place. You are a node on an energy network. The Railroad Commission doesn't see a town. They see what Martin Heidegger called a "standing reserve," a stockpile of resources to be mobilized and used up.[27]

From the state's perspective of running a complex technological system, the informed consent of ignorant amateurs who happen to live in proximity to it is irrelevant. It won't improve the functioning of the system any more than getting the consent of research subjects will improve the validity of a study's conclusions. Indeed, in both cases the requirement of informed consent can throw a major wrench into the works. Some of the Nazi experiments (for example, those on hypothermia) were sound science, but they involved such pain that they would never have been run if people had the choice to opt out of them. What this means is that if we value autonomy highly, then some experiments (whether on the shale world or in the clinic) just won't happen.

The state government is not in the business of getting the consent of people living near fracking sites. To involve the consent of the vulnerable in the politics of fracking, then, we need a different kind of public sphere. It would have to be one that is oriented not toward system functionality (getting the gas out of the ground efficiently) but toward protection of goods like health, safety, beauty, and community integrity—those broader values that might be unintentionally sacrificed in the name of a narrowly framed goal like producing gas at a profit. In the case of fracking, that public sphere is town and municipal governments.

Local government is, in other words, the political home of informed consent. It is the voice of those living on the surface and made vulnerable to the harms caused by real-world experiments. Admittedly, we don't usually think of this in terms of consent. But whether it is community rights, public liberty, local control, or participatory justice, we are talking about the same core principle, namely, that you deserve a say over decisions that importantly impact your life.

This is why municipalities have become the most important flash point in the politics of fracking: they represent a different moral order, one that is rooted in place and community rather than commodity production. This is the threat posed by local resistance that has the industry nervous enough to use psy-ops. Their business model simply doesn't work if those bearing the risks are empowered to dictate the terms of operation. For those who are making millions from the newly viable reserves, the focus has shifted from the technical question of how to get the gas flowing to the political question of how to keep it that way. And they are struggling (despite friendly state agencies[28] and federal policies that embrace fracking as part of an "all of the above" energy policy) because across the country local groups like DAG are hounding them.

Conflicts between municipalities and oil and gas development are nothing new.[29] The epicenter is probably Los Angeles, which sits atop several productive oil fields. There are amazing pictures from the turn of the century of a thick, bristly forest of oil derricks engulfing neighborhoods. And to this day, oil operations—often hidden by the facade of a fake building—are at work in the heart of Los Angeles.

In 1965, Huber Smutz, the chief zoning administrator for the City of Los Angeles, wrote a biting assessment of urban drilling and the "totally selfish [and] thoughtless" acts of oil companies. If the industry wanted to drill in LA, they had to go through his office to get a permit. "We would prefer that oil drilling and production be totally excluded from urban areas, particularly residential sections," Smutz said, but he mournfully noted, "Sometimes such exclusion would be unreasonable and thus legally impossible." Therefore, he continued, "It has become the planner's duty . . . to make sure that if oil drilling is permitted in such communities, it is allowed only under strict control." [30]

Throughout the mid-twentieth century, a steady stream of court cases confirmed the legal authority of municipalities to exert some measure of control over oil and gas development.[31] In 1933, the US Supreme Court upheld an Oklahoma City ordinance requiring drillers to hold a $200,000 bond prior to drilling a well within the city's jurisdiction. The first legal test of a Texas municipality's authority came when Tysco Oil Company sued South Houston in 1935. In a

terse opinion, the district court judge rejected Tysco's assertion that
South Houston's ordinance was "arbitrary and unreasonable."

The next landmark case was the 1944 *Klepak v. Humble Oil and
Refining Co.*, in which the Galveston Court of Civil Appeals argued
that the Railroad Commission's authority did not subsume existing
police powers vested in municipalities to also regulate oil and gas
activities as long as municipal rules are not "unreasonable, arbitrary,
[or] discriminatory."

In 1982, just as Mitchell began tapping into the Barnett outside
of Fort Worth, the Fort Worth Court of Appeals further bolstered
this opinion. In *Unger v. State*, they rejected the argument that the
Railroad Commission wholly preempted municipalities from promul-
gating their own regulations. They also, for the first time, conceived of
municipal oil and gas ordinances as zoning laws. This opened a new
line of reasoning to the effect that municipalities can legitimately regulate oil and gas activities just as they regulate any other kind of land
use. Twenty years later, as the fracking boom came to the Dallas–Fort
Worth metroplex, the next round of feuds began between cities and
the industry.

Bryn Meredith, the attorney for the wealthy city of Southlake just
south of Denton, has been in the middle of those fights for over a
decade. He is widely regarded as one of the top three experts on municipal oil and gas law on the Barnett Shale. He is a slender, gentle man
with a young face, kind eyes, and a predilection for Starbucks. One
rainy day while DAG was still mulling over the question of whether
to pursue a ban, I met Bryn in his spacious Fort Worth office. In the
waiting area was a bronze statue of an Indian hunting a pair of bison.
One bison had a spear lodged in its back; the other was flaring its nos-
trils. I tried to comprehend how we got from that world to the maze of
highways outside the window.

Bryn graduated from law school just as the Barnett boom began.
He first worked for Newark and Reno, both Texas cities. In 2002,

Devon applied for a permit to drill in Reno and told the city it had no jurisdiction over the matter. "But they were either ignorant or lying," Bryn told me. "Oil and gas development is a surface land use that is subject to municipal regulation."

Reno is northwest of Fort Worth and has been transformed recently from a small farm town into a bedroom commuter suburb. In 2003, Reno City Council was about to deny a permit for Devon to drill a well in a residential neighborhood. Devon hired a law firm and brought a court stenographer to the next public hearing to intimidate City Council. They were getting ready to sue.

Reno, unlike Southlake, had a minuscule tax base, and its insurer would not cover a lawsuit. Devon is a Fortune 500 company with cash inflows of $10 billion and twice as many employees as the population of Reno. If Reno had lost in court it would have been costly. City Council yielded, approving the permit with a few minor concessions.

Bryn has ever since been learning how to effectively use every ounce of municipal power to protect cities from oil and gas development. He is also the lawyer for Dish, and he fought with Calvin Tillman to fend off the pipeline industry there. Much is made about Calvin's battle since it was featured in *Gasland*, but few people realize that Bryn, this unassuming man in a maroon sweater, was his legal field marshal.

During our visit, Bryn recalled a recent meeting in Dish when an angry resident looked at him and assumed he was an industry plant. He yelled, "I'm going to call Calvin Tillman—he'll send us someone to represent the people." As cool as he could be, Bryn replied, "You go ahead and call Calvin. I've orchestrated every campaign he's ever launched against the industry." To his credit, the man apologized for assuming all attorneys are the enemy.

I didn't tell Bryn about how we were pondering a ban, but I did ask him what he thought about the dozens of municipal fracking bans around the country. "I understand the sentiment behind them," he said, "but they are not a very artful use of municipal power." He didn't directly say it, but one could infer a message like this: it's better to write an ordinance that "permits" fracking but is in fact so robust as to constitute a de facto ban. Many felt that this was precisely what he had so masterfully accomplished at Southlake.

During the Frack Free Denton campaign, the question of local authority was very much in the air as courts across the country continued the case-law tradition that had begun in the 1930s. In June 2014, the town of Dryden, New York, delivered a knockout blow to the industry when the state's highest court upheld its ban on fracking.[32] The court agreed with an earlier ruling that the ban is not preempted by state law; it does not regulate the industry so much as do what local laws should be able to do, namely, protect neighborhoods and community character. The Pennsylvania State Supreme Court made a similar argument around the same time when it struck down the portions of Act 13 (a law governing oil and gas development) that had essentially neutered all municipal authority.[33]

But in July, when the Briggle family was back in Colorado on vacation, we got the bad news via a text message from Sharon that a judge in Boulder County had made an initial ruling against Longmont's fracking ban.[34] Shortly after that another Colorado judge struck down Loveland's five-year moratorium. The opposite rulings in New York and Colorado made it clear that the situation was state specific. We felt strongly our ordinance was defensible, but there was no way to know.

One thing was certain, though: the industry was going to try to convince Denton voters that the ban was illegal and would sink the city in lawsuits. If the ban wasn't preempted by state law, they said, then it would certainly constitute a regulatory taking of mineral owners' private property. Even Dalton, much to my chagrin, focused on the potential legal costs. At one point he went on camera and said the costs would be "tens and hundreds of millions. It would bankrupt the city."[35]

While fracking is a relatively new issue, disagreements over municipal power have deep roots reaching back to the tumult and bloodshed of seventeenth-century Europe. That's when Thomas Hobbes argued that we must have a Leviathan, a single ruler, to quash disagreements and ensure peace. For Hobbes, the thought of a bunch of cities making their own rules was anathema. It would lead to chaos and conflict. He lamented "the immoderate greatness" of towns in the commonwealth and saw cities as deviants, lesser communities in the bowels of a greater community like "wormes in the entrayles of a naturall man." They are forever "Disputing against absolute Power . . . [and] medling with the Fundamentall Lawes."[36] Hobbes would have been a good Texas railroad commissioner.

Until the nineteenth century, cities were just like any other mercantile or commercial entity. The public/private divide didn't exist, at least not in its current form. Corporations, so conceived, were a remnant of feudalism, a messy intermediary between the modern poles of individual and state. Gradually, through early nineteenth-century US court cases, the categories were purified. "Private" corporations gravitated to the individual pole (the free market) and were protected from state domination. "Public" corporations (i.e., cities) gravitated to the state pole and were subject to its central authority.

The industrialization of America in the late nineteenth century opened a new chapter in the now tripartite battle between state, city, and corporation. State limitations on cities were widely invoked through arguments that local regulation of business constituted interference in the private sphere. The federal judge John Forrest Dillon nearly killed city autonomy in his 1872 legal treatise titled *Municipal Corporations*. According to "Dillon's Rule," cities "owe their origin to, and derive their powers and rights wholly from, the legislature. It breathes into them the breath of life, without which they cannot exist. As it creates, so may it destroy."[37]

The defenders of cities fought back with arguments about the democratic function of local politics and the absolute right of local self-government. It was during this era that two crucial weapons for city power were invented. First, several state constitutions were amended to grant certain cities the power of "home rule," meaning the ability to enact legislation without explicit state permission and the ability to prevent some state invasions on local autonomy. Denton is a home rule city. Second, the *Euclid* Supreme Court case of 1926 validated zoning as a permissible exercise of municipal authority to restrict certain uses of private property.

Despite such efforts, just as private corporations were being upgraded to "persons" with constitutionally protected rights, municipal corporations were being downgraded to "creatures of the state" with no such rights. And whereas private corporations have since clearly grown in influence over politics (witness *Citizens United*), cities have been able to cobble together only a limited and ambiguous set of powers. Local resistance to fracking, like the Occupy movement, is responding to this deeper current of history. It is a rejection of the centralization of state power, corporate influence on state leadership,[38] and the increasingly untenable notion that corporations are persons acting solely in a politically neutral private sphere that we call the free market.

Of course, I don't want to romanticize local government. Through DAG's earlier advisory work on the gas well ordinance, I had seen its ugly underside. In an age of dependence on advanced technology, even local government slides into a systems technocracy—it has all the trappings of a democracy, but in reality the gravitational center of power is the unelected bureaucracy managing the unelected machines. Still, at the local level it's easier for an average citizen to plug into that power and give it a nudge this way or that. Alexis de Tocqueville wrote that local government is the crucible that forges citizens out of consumers. Despite their shortcomings, cities and towns still bring political liberty—the chance to act as citizen and not mere consumer—within people's reach.

The ban was part of this long history of a jurisdictional tug-of-war between cities, states, and private corporations. Up to that point, Denton had been losing the war, getting dragged through the mud by the bigger kids on the other end of the rope. City Council, for example, couldn't stop Range Resources from fracking their wells at the purple park. All they could do was order them to build a stone wall around the site. Whenever I rode my bike past there along Bonnie Brae I thought of that wall as a symbol of community powerlessness. Wells would be fracked. All we could do was hide that fact, not prevent it. But maybe we were about to change all that.

CHAPTER 6

IN THE
BALANCE

S THE CAMPAIGN TO BAN FRACKING GAINED
momentum across the spring, the industry and its supporters (including some mineral rights owners associations) started ramping up their attacks against us. At first, the opposition took the form of a "grassroots organization" (i.e., industry astroturf) called North Texans for Natural Gas, which sprang up shortly after we announced the ban and somehow racked up 10,000 Likes on its Facebook page in the first few weeks. They were all over Facebook with memes saying, "Like us if you support American energy independence." [1]

However, most of the campaign against the ban was run by a PAC called Denton Taxpayers for a Strong Economy. The website listed two board members, Bobby Jones and Randy Sorrells, both Denton residents. Together, they owned more mineral wealth in Denton than the entire Denton Independent School District. They hired the high-powered and highly decorated Fort Worth–based Eppstein Group to handle their political campaign. The Eppstein Group boasts a 90 percent winning record across 1,500 elections. I imagined people in designer suits sitting around a mahogany conference table in a downtown skyscraper. Meanwhile, we had coffee at Rhonda's house scheduled around Cathy's appointments with patients and my kids' bath time.

I'll get to the various strategies (including Russian conspiracies) that the industry used to attack us. But for now, I'll focus on an angle most often used by individual trolls and detractors on our Facebook page. I want to do so because it pushes us to rethink this whole idea of informed consent and the ethics of real-world experiments.

Here's the argument: Frack Free Denton is nothing more than NIMBYism—just a bunch of entitled suburbanites screaming, "Not in my back yard!" I first encountered this argument when Cathy, Rhonda, and I were handing out flyers in the Meadows back in September 2013. I left a flyer on one gentleman's door, and he wrote an e-mail back: "All the stuff you have is because of this industry [oil and gas]. If you don't like it, you should stop driving or using anything related to it." Another typical comment went something like "You should all ride bicycles and burn wood. No cars or electricity for you!"

The problem with the analogy to clinical trials, the argument goes, is that we aren't just guinea pigs of the shale but also consumers of the shale. Oil and gas development makes our way of life possible, so it is hypocritical to criticize or reject it. You could even use Kant to make this point, because he said you must never make an exception of yourself. That's what the NIMBY critique is claiming: you can't consume the goods made possible by natural gas while refusing the downsides, because you are in effect saying that someone else should deal with the downsides. But why do you get a free pass? What makes you more special than that other person? Nothing, Kant would say, because we all possess equal worth and dignity.

It was around this time that the Briggle family started our brief and ill-fated experiment with family grocery shopping trips. Instead of Amber going out on her own, all four of us would go. MG did a pretty good job staying on task, but Lulu was a whole other story. Everything was "oooo . . . pretty awesome!" and needed to come off the shelves, be carried around briefly, and then be redistributed on the floor. After a while, I got burned out restocking the entire store and decided to stay home with Lulu on shopping days.

All those cans and packages that Lulu would grab are perfect symbols of our modern condition. So is the air conditioner that we started turning on again as April rolled into May. To be modern is to be surrounded by commodities made readily available in a foreground of consumption. Instant food. Instant cold air. To be modern is also supposed to mean that we are sheltered from the background of production with the dirty machines that make it all possible.[2] I can get a pound of ground beef wrapped in cellophane and I don't have to know anything about the messy process that got it there. The same thing holds for the energy infrastructure that brings us cold air in the summer.

What's happening with fracking is that the foreground of our daily lives is being invaded by the background machinery that makes those lives possible.[3] Those who called us NIMBYs were saying that by consuming the commodities in the foreground we had implicitly given our consent to the background. To eat the hamburger is to agree to the slaughterhouse. To cool the home and drive the car is to agree to the gas and oil wells. If you don't like it when the background conditions are shoved in your face, then stop participating in the foreground of consumption. It's kind of like the biblical saying "Let he who is without sin cast the first stone."

There is something right about this argument. I often wondered how devoting my days to a fracking ban could possibly square with my evenings when I would draw a hot bath for the kids and listen to the natural gas-fired water heater kick on in the closet next to the bathroom. It seemed to me that Ghandi's response was partially correct: I had to be the change I wished to see in the world. In fact, that summer Amber and I poured as much money as we could into making our home more energy efficient, including installing new windows.

Acting as individual consumers, though, we can only make changes

within the system. We can't make changes *to* the system, that is, the way
the modern world is structured to foster lives of utter dependence on
fossil fuels. We can't just, as my letter writer suggested, stop using stuff,
at least not in one fell swoop. Jean-Jacques Rousseau, Kant's inspiration, wrote about this in terms of humanity inadvertently imposing a
yoke on itself as it pursues conveniences and comforts through technology. At some point they transmogrify from mere desires to *needs*
that we can't really do without.[4] In other words, we are not as autonomous as the NIMBY critique seems to presuppose.

If only sinless angels could lobby for social change, then we'd never
get anywhere. If we couldn't reject something just because we've grown
dependent on it, then we'd still be insulating our schools with asbestos. Given our situation of dependence, I think the NIMBY critique is
like telling a woman who is financially dependent on an abusive husband that she can either put up with the abuse or drop everything and
leave him overnight. That's a double bind and a false dilemma.

Plus, there is something hypocritical and privileged hidden in the
NIMBY critique, because if you have enough money, then you don't
ever have to put up with the messy background of production. You can
just move or use your clout to keep the polluting factories far away. I
suppose if you had enough money you might even be able to purchase
the moral purity that purveyors of the NIMBY critique claim you
must have before you criticize oil and gas development. The trouble is
that the 99 percent can't afford to first buy a new electric car and then
protest the gas well outside their kids' bedrooms.

Further, thinking like the proactionary Julian Simon, we can say
that NIMBYism is not about making an exception for one's self.
Rather, it's about expressing preferences and using the market to
respond. If people refuse to allow fracking in their back yards and this
causes supply to drop, then prices will rise, which will create incentives
to find alternative ways of satisfying demands, in this case for energy.
We could NIMBY ourselves into an alternative energy future!

The more I think about this criticism, the more absurd I find it. Almost every city in the world doesn't have fracking in its borders; does that mean all their residents are hypocrites for not having frack sites in their back yards? How does the geological accident of sitting atop a shale reserve suddenly amount to an imperative to frack away recklessly right next to homes? Or as Cathy put it once when she was being heckled at a public event: "Since when did it become part of the deal that using electricity means we have to poison our kids?"

Besides, all of this is fighting on the wrong turf. It presumes that the correct framing here is about energy policy. Frack Free Denton wasn't about that, though. It was about land use policy and community integrity. And in that realm, our society widely acknowledges "Not in my back yard" as a valid ethical position. It's the logic behind zoning, which is used in countless cities in the country. There is an appropriate place for each kind of land use and some of them do not mix. City Council member Jim Engelbrecht always delighted in asking natural gas industry representatives at City Hall if they would like to have a frack site in their back yard. No one ever said yes.

The only reason we had the insanity of fracking in back yards was the predominance of the mineral estate and the production-mad mentality that goes along with it. Back in 2012 when we were revising the ordinance, we would have even settled for the 1,200-foot setback distance, but the industry insisted on claiming their right to frack less than 200 feet from homes. They tried to label us as radicals and extremists, but we only wanted commonsense neighborhood protections. They were the extremists. And, with their unneighborly ways, they brought the ban on their own heads.

BACK TO THE clinical trials analogy. There is a distinction in medicine between "therapy" and "research." The term *therapy* typically applies to practices that are intended to promote the health and well-being of the patient. *Research*, by contrast, is an activity designed to test a hypothesis and contribute to generalizable knowledge. When you are a patient of a therapy, the goal is to benefit you. When you are a participant in a research trial, the goal is to use you to benefit others. The only ethical way to *use* people like this is to get their informed consent. I argued that this should apply to real-world experiments too. As one scholar put it, "No innovation without representation."[5]

What this NIMBY debate shows is how the border between research and therapy is fuzzy. In the case of fracking, the people directly suffering the harms of the natural gas and oil "research" in proximity to their homes also indirectly reap the benefits of the "therapy" as consumers of oil and gas. They may not get gas (or electricity or chemicals) from the particular well near them, but they benefit from the network that that well feeds into. That's what, supposedly, makes one hypocritical for saying "Not in my back yard."

But think about NIMBS, or "Not in my bloodstream" (a label I made up, but which probably suits most Americans, judging from the fact that clinical trials for drugs intended for American consumers are increasingly being conducted on poor people overseas[6]). Those who refuse to take part in biomedical research do not face the same accusation of hypocrisy. Yet they too indirectly benefit in the same way. They may not be helped by the drug they are testing, but they benefit from the network of medicine that that drug feeds into. And that medicine was derived from past experiments on human participants.

So, do we owe a debt to society to take some risks by living next to gas wells or ingesting experimental drugs? After all, that's how society progresses. As Max More noted, every innovation will create desirable and undesirable effects—some people are hurt along the way, but through their sacrifices things get better and better.[7]

Yet I agree with the philosopher Hans Jonas who said we, as individuals, are not obligated to advance the progress of science and technology in the name of the greater good. Writing about medical experimentation, he said that we are indeed indebted to past "martyrs" who risked their safety to advance the medical knowledge that we now enjoy. But society is also indebted to those martyrs, "and society has no right to call in my personal debt by way of adding new to its own."[8]

For Jonas, an emergency is the only time society may override the individual's right to consent—as when a draft is implemented in times of war. But a slowed rate of medical progress is not an emergency. Though individuals will, of course, continue to die, society will go on even if cancer and heart disease are never cured. A slowed rate of fossil fuel development is also not an emergency. Human societies survived for centuries before the age of oil and gas, and alternative energy sources are increasingly viable. If anything, as climate change shifts from a future likelihood to a present reality, continued fossil fuel dependence is far more likely to precipitate than to prevent an emergency.

Even on an individual level, we might well fare better if we were to consume less energy—it's not clear that Americans are happier now than we were sixty years ago when we used half as much energy and consumed half as much stuff.[9] More energy is not always better; just look at the frenetic pace of our overworked lives in an age when "energy slaves" should have long ago freed us all to pursue our passions. Japan scores just as high on indices of well-being as America, but an average Japanese consumes half as much energy as an average American.[10] That point is sorely missing from the fracking debate. Of course we should focus on the environmental and health hazards of the background machinery. But the problem isn't just the recklessness of the means; it's also the mindlessness of the ends.[11]

It's true that progress in science and technology often requires some to shoulder a disproportionate share of the burden. Who should pay the price—should it be Maile and her children? Jonas argues that sacrifices in the name of science should only be made by those who choose to take the risk because they genuinely identify with the cause at stake. Thus, scientists themselves should be the first in line to volunteer, and indeed there is a noble but now largely moribund tradition of self-experimentation in the sciences. Next in line would be patients suffering from the disease the drug is intended to target. Then maybe those patients' families. At the bottom of the list would be poor children in India who do not even understand the trial.

The same kind of ranking should hold for fracking and other hazardous technological activities. I think that's what got people so riled up when Rex Tillerson, the CEO of ExxonMobil, joined a lawsuit protesting the construction of a water tower (that would be used in part to supply fracking sites) near his home in Bartonville, just south of Denton.[12] He cited concerns that the nearby industrial activity would devalue his property. If *anyone* should live with the mess that fracking creates, it should be the guy who makes $40 million per year putting those messes in other people's back yards. If he's not willing to consent to it, then why should it be forced upon neighborhoods like Maile's?

What Jonas warned about was "the erosion of moral values" that happens when we start thinking about scientific or material progress as an unimpeachable good. Recall what the captain, full of Faustian verve, said to Frankenstein: "One man's life or death were but a small price to pay for the acquirement of the knowledge which I sought." That's the danger of the NIMBY critique. If we are not careful, it mutates into a ruthless utilitarianism where a mom's peace of mind and the health of her children are on the losing side of the ledger as we tally up the greatest good for the greatest number. We must never think of people's lives as small prices to pay.

WE ONLY HAD to get about 600 signatures to put our proposed ban on the ballot as an ordinance. We probably got close to 200 signatures at the petition launch party in February. Over the next few days we took the petition to Fuzzy's Tacos and a Mardi Gras party downtown. It was hard to keep track of all the petitioners. Some of them were excited to be involved at the big launch party, but then their commitment must have waned, because they took their clipboards and petitions and we never heard from them again. This drove Cathy crazy. Most petitioners, however, were working diligently away, and I was pretty confident we exceeded our requirement within the first two weeks.

By law, we had 180 days to get the required signatures. We could easily beat that, which raised the question of timing: When should we turn the petition in to City Hall? After we turned it in, City Council would have 60 days to schedule a public hearing where they could either accept the ordinance as proposed or put it on the ballot for a citywide vote. We were sure City Council would send it to the voters, but we also realized we were too late to get it on the local ballot in May 2014.

Initially, we thought that the city charter stipulated that such situations called for a special election. That seemed favorable to us, because only voters who cared about this issue would come to the polls. As it turns out, however, we had misread the legal language, leaving us with two choices, neither one ideal. We could time things to get us on the November 2014 ballot, or we could stretch it out just long enough to miss that window and land us on the May 2015 ballot.

I pushed for the latter option, because of the way the numbers worked out. In state and national November elections, Denton sees about 25,000 voters hit the polls. In local May elections, that number is just 5,000. Far easier to get 2,500 votes than 12,500, I reasoned. In addition, the November midterm election would at least implicitly frame the ban not in the context of local land use policy and neighborhood industrialization but in the context of ideologically charged national politics. All the industry would have to do was paint the ban as part of a "liberal agenda" and the straight-ticket voters would reject it after ticking the box for Greg Abbott, the Republican nominee for governor. Given that Denton basically tracks with the rest of Texas in voting Republican, we'd be sunk.

Ultimately, though, we decided to shoot for November, because we thought the attrition of dragging the campaign across fifteen months would be even worse. We couldn't see how we could build momentum and hold people's attention all the way to the following May. Besides, we were really working in the dark—we couldn't afford polling at that time, so we didn't know which option would be best. Sure, we were a PAC, but we were no super PAC!

The majority vote is also a kind of consent. In fact, it's the way democratic societies usually express consent, or "the will of the people." The cold reality of the vote on the ban forced me to reexamine the analogy between fracking and clinical trials. In clinical trials we are only talking about the consent of one person (that is, each person individually chooses to either sign up for the trial or not), but with fracking (or, really, any industrial siting issue) we would need the unanimous consent of dozens or even thousands of people in the surrounding area. In the former case we are dealing with the human body; in the latter case we are dealing with the political body. Can we really talk about consent when it comes to a collective?

Supposedly, what makes a society democratic is that it is sustained by the consent of the governed. But "the people" disagree with one another. Locke imagined that originally *"every* individual" consented to join a community, making it into "one body." But then decisions have to get made. The communal body can only move one way or the other, even though its factious individual elements want to push it every which way. Locke said that "it is necessary the body should move that way whither the greater force carries it, which is the consent of the majority." [13]

Majority rule may be a practical necessity of communal life, but it always bothered me that the ban came down to it. According to my first criterion of ethical innovation, those most vulnerable to harm should have the greatest say. That would mean something like neighborhood-level governance, but we didn't have that option. The smallest level of government we had was municipal.

We think of democratic government as the referee of the marketplace, not picking any sides but simply ensuring that we are all free to choose from an ever-expanding array of technological means to achieve whatever ends we want. More technology means expanded opportunities for people to shape their destiny.[14]

But this picture doesn't compare well with reality. We are not free to refuse cars and highways or the Internet and smart phones. At least, we are not able to escape the substantial changes these technologies bring about in our lives. One can, of course, choose to not own a smart phone (I held out for years). But as society builds expectations and opportunities around these devices, those who do not have them are marginalized. Modern technologies have a way of transforming desires into needs and private preferences into public realities—realities that one is not simply free to accept or decline. Millions of residents driving automobiles in big cities produce smog that no one is free to ignore.

We can't prevent our actions from impacting others or their actions from impacting us. This is especially so in an age where technology magnifies our reach. Technology might give us the illusion of isolation (with our big SUVs and home entertainment systems), but by weaving our fates more closely together, it's creating a reality of unprecedented interdependence. You might even argue that individual rights are anachronistic in an age of high technology: an increasing population of individuals feeling entitled to consume as much as they want is a sure ticket to planetary doom.[15]

This is the problem with the NIMBY critics who say we should just stop consuming fossil fuels. We are not heroically the sole authors of our world.[16] We can't just wish away the car-centrism of the metroplex. How are we going to get to work? Of course, this point cuts both ways and could undermine my defense of a consent-based NIMBYism. Given that our lives are so intertwined, there's no possible way each of us could be asked to consent to everything that will impact us. What, we'd have to get consent from everyone on the street to do some road or water-line repair? Would I need to get my neighbor to sign off before I run my lawn mower? It sounds like another way that a precautionary mindset brings the world to a grinding halt.

Does this mean that the ideal of informed consent for real-world experiments is utopian? In one sense, yes. We are thrown into a world the features of which we did not choose and largely cannot control. I think this is why Max More, in his own formulation of the proaction-ary principle, talked about compensation rather than informed con-sent: it's foolish to think we can control how innovation will unfurl. It's not so foolish, he figured, to expect some compensation (after the fact) if we get squashed by the wheels of progress.

More might argue that informed consent is a precautionary norm, because it involves a gatekeeping mechanism *prior to* action. But we need to be careful about what "action" means. It is true that Cathy, Sharon, and Maile could not consent to the first natural gas well ever drilled or to the development of fracking or to the wider fossil-fuel-dependent world in which they find themselves. These big actions are beyond their control. But they certainly *could* give or refuse their consent to the *specific* fracking projects in their neighborhoods, that is, if the political arena were structured so as to afford them that chance. This is a small action in medias res— in the context of the ongoing tumult of global innovation. It is an opportunity to protect and exercise the central proactionary and democratic value of freedom.

The trouble at the purple park, at the Meadows neighborhood, and in hundreds of cases like them around the country is not that informed consent is an unrealistic goal. It's that the political frame-work is not designed to empower those who should have the greatest say in the matter. What might an alternative political ontology—a different kind of sandbox—look like? To answer that question, I needed the help of one of my students.

IN MAY, an unidentified group started calling Denton residents. They were conducting a poll about the ban. I never got the call because I don't have a landline, but Ed did. He was sitting in his living room, surrounded by his collection of rare books, petting his little dog Willie when the phone rang. "How do you feel about a group of liberal college students, who do not live in Denton, pushing their agenda for a fracking ban on your town?" and "Would it change your opinion if you knew the ban would make Denton one of the most over-regulated communities in the country—on the same level as France?" "Would you be more likely to vote against the ban if you knew it was being funded by Putin and Washington, DC, liberals?" They then went on to ask Ed's opinions about various political figures from President Obama, Ted Cruz, and Dick Armey on down to Denton City Council member Kevin Roden and even "UNT professor Adam Briggle." The whole experience was so bizarre that Ed nearly fell off the couch.

It was right around that time that I finally lined up a driving tour with my former student Matt Story. Matt had recently graduated with a master's degree in philosophy from UNT, and along the way he had taken my course "The Philosophy of Energy Policy." During one of my lectures about the municipal control of fracking, he had the usual enigmatic look on his face. So I asked him, "What are you thinking?"

"With all due respect," he replied, "you don't know what you are talking about." Matt grew up about fifteen miles west of Denton, in the heart of the gas patch. "You should come to my old house," he said, "if you want to see the reality."

Matt is a soft-spoken man with a buzz cut, bright blue eyes, and a sly, youthful smile. When we used to talk about the philosophy of gas wells in class, his eyes would beam and his mouth would slant as if he knew the answer to the questions I wasn't even thinking to ask. So I figured he might help me get some clarity—or at least put me into a more interesting muddle.

On a cloudless day, Matt picked me up at my house in his white SUV. We set out for a half day's meandering tour through points west of town, deep into the origins of the Barnett boom. "I guess you could say that I'm puzzled by the fact that the city folks in Denton got so outraged about gas wells within a few hundred feet of homes," Matt remarked. "Almost everyone in rural areas outside of town has at least one gas well at least that close to their house." He went on to tell me that most of those families made some money from surface use agreements with the operators to cover damages and inconveniences, but hardly anyone was striking it rich. In fact, Matt guessed that only 20 percent of folks in the rural parts west of town owned their mineral rights. "My family never owned the minerals. I still don't know who collects the royalty checks for the gas under our old house."

We drove past the city limits into a land of hay and brush. "Everyone out here has a story to trade about spills or accidents, but no one really puts up a fuss. It's just part of life," Matt explained with a shrug of his shoulders. I understood him to mean that fracking is just another thing to trudge through in the quiet revelry of Texan self-reliance. Droughts happen. Gas wells happen. So it goes.

Denton is near the 100th meridian, the line that symbolizes the beginning of the American West. It's where things start getting too arid to raise crops. To the east, savannas give way to oak forests and eventually bayous. But to the west, near Matt's old house, lonely post oaks fade into parched prairies of mesquite and buffalo grass.

We traveled along farm-to-market roads in what Matt calls "the balance." It's all that wide open space (over 90 percent of the land area in Texas) that is not incorporated into a city, at least not a city big enough to have home rule powers. I wanted to understand what Matt thought I was missing, so I asked, "What's your take on the politics of fracking?"

He thought for a moment, the sun glinting on his shades as a field of green wheat slipped by outside the window. "Let's put it this way," he said. "The people in Denton are citizens of the city. But out here, they are citizens of the Railroad Commission." Then he got more animated than usual. "But that makes no sense, because the Railroad Commission is in the business of extracting minerals, not recognizing citizenship!"

That's why gas wells are so close to homes out in the balance. The Railroad Commission does not regulate setback distances, because its main concern is with the mineral riches below, not the surface impacts above. Proximity is governed by the lowest common denominator, the international fire code, which allows gas wells to be as close as 100 feet from a home (just enough room to maneuver a fire truck). It takes a city to impose setback distances greater than that.

We drove by dozens of little houses isolated on large tracts with a few goats or cattle, just enough to retain their tax status as agricultural land. At one house, a boy was playing basketball in the driveway. He could have easily thrown the ball onto a neighboring gas well site. We drove through tiny towns like Krum, Ponder, Forestburg, and Dish. The high school boys out there still play football, that greatest of Texan religions, but enrollments are so small that when the Friday night lights illuminate the gridiron each team has only six players on the field.

We were surrounded by gas wells. "You can date the sites by the color of the paint on the condensate tanks," Matt told me. "The oldest equipment that has been around for at least twenty years is rusty metal with no paint. The next generation is a faded green. Then there is a generation of maroon. The newest stuff looks like that," he said, pointing to a tank painted bright white.

When Matt told me everyone in the balance is on well water, I asked him if there were stories of water contamination from fracking. He just shrugged his shoulders. "I'm not sure about around here, but my parents' well has been totally ruined."

His parents had moved north of there to escape all the fracking. But shortly after they settled into their new house, a compressor station (a set of engines that push the gas down the pipelines) was set up just outside their front gate. After drilling activity occurred around there, their water well got so polluted that they had to invest in a complex filtration system. "The water smells like rotten eggs," Matt said, "and if you put some of it out in the sun and let it evaporate, it leaves behind a bunch of calcified sediment at the bottom of the glass."

I had heard lots of stories of folks seeking a quiet life in the country only to be chased out by the industrialization that came with the Barnett boom. Refugees of the shale. The rancher who supplied our local beef, for example, had two wells and two pipelines built on his property. Another five wells with compressors are within a quarter mile. "My ranch used to be peaceful," he once told me. "Now it's loud and unpleasant. Not much fun to be there anymore."

We drove on, watching the occasional Bridgeport Tank Truck carrying fracking wastewater over the horizon through mile after mile of low rolling hills incised by dry creek beds. I could see farmers using skid loaders to dig up piles of dirt, and I wondered at what depth they became trespassers on the mineral estate.

Matt's father built their old house in the 1960s. It has white walls, a green roof, and, of course, a gas well just outside the front door. When we arrived there, we pulled off to the side of the road. Matt and I chatted for a moment about field philosophy. I told him how much I had been agonizing about my shift from policy advisor in the early years of DAG to advocate for a ban with Frack Free Denton. This wasn't a matter of Socratic questioning and doubt. I asked Matt, "Is this philosophy?"

He shrugged, smiled, and opened his eyes real wide. "Beats me, man!"

I told him how I had come to terms with my new role. No single individual needs to speak for every side of an issue.[17] If we each do our part, speak from our perspective and convictions, it adds up to a marketplace of ideas. Maybe the field philosopher's main obligation is to steward that marketplace—to make sure that it remains diverse and robust. The philosopher Karl Popper argued that there is a constant threat that powerful ways of thinking will homogenize the marketplace of ideas and reign unchallenged even though they may not be true or right or just. What characterizes both science and democracy is a tireless vigilance of critical reasoning that questions ruling theories and policies—puts them constantly to the test.

In that vein, we can see the field philosopher's task as picking out an idea that's been given short shrift—that's deviant or an underdog— and championing it, making the strongest possible case for it. For me, the idea is about local self-determination in the face of powerful state and corporate interests. Understood this way, doubt and questioning are far less important than sheer doggedness. You must take your marginalized position, forge it into an ax, and strike the mightiest blows you can muster at the powers that be. For if you don't do this, how will we ever know for sure if our widely held beliefs and structures of authority are really the right ones?

So the field philosopher can be like Socrates in the agora paralyzing people by making them see how all their actions are based on unexamined assumptions. But he or she could also be like the French philosopher Michel Foucault marching in the streets for the rights of prisoners.

"Either way," I said to Matt, "we are not talking about philosopher kings. It's more like philosopher bureaucrats, because the philosopher is at work in democratic processes—"

"But do you call this democracy?" Matt asked, cutting me off and gesturing at the gas well near his old home. "The problem in the balance is that there are no conditions for politics. There is no platform on which to stand and be recognized as citizens. That's the difference with the city. If you live there and you have a problem, you go to City Hall. Out here if you have a problem you go to the liquor store and your front porch. Maybe you get your neighbors to join you. But even gathered together, you don't comprise a political institution. You don't form an entity that anyone has to answer to."

ON THE DRIVE BACK, Matt counted all the gas wells along the road between his old house and Denton. It had been a while since he had been out there, so he wanted to see how many new ones had been drilled.

While he counted, I kept thinking about what he had said. Under the existing regime, an operator that wants to develop a well must get the appropriate federal, state, and local permits. Although these laws and permits change, they are anchored to stable political entities. That's what I hadn't seen until Matt pointed it out to me. What he was saying was that for my ideas about informed consent to really work, any time an operator wants to frack a well, an entirely novel political unit—one that is ephemeral and hyperlocal—would need to emerge. My thinking was stuck in the existing political sandboxes (federal, state, and municipal). Matt got me thinking about the possibility of political power even more local, and more transient, than cities.

Let's call it a deme in honor of ancient Greece, the first democracy, and the way Cleisthenes imagined citizenship as rooted in place rather than in genes. The ancient demes were essentially the first voting districts. Unlike a voting district, however, the demes I have in mind would not have predetermined parameters. Their boundaries would depend on where the activity is proposed. Let's say anyone living or working within a half-mile radius around the proposed frack site would band together, forming a new and temporary deme empowered with the authority to decide whether and under what conditions fracking could occur. Everyone in this radius would participate, regardless of their status as mineral or surface property owners.

The half-mile radius is somewhat arbitrary (though it is informed by scientific studies of the range of likely harms and could be adjusted as we learn more), but it is anchored in the general ideal of participation based on proximity.[18] It is less arbitrary than the historical accident of mineral ownership and even the municipal vote that gave everyone in town equal say regardless of their proximity to fracking.

Matt broke the silence to give an update on the gas well count. "That's thirty," he said. We had been driving for five minutes.

The deme sounds impractical, even though the economist Elinor Ostrom showed that localized decision-making communities are widespread.[19] Yet Ostrom was talking about premodern practices like ranching and fishing where the resource being managed fits the parameters of the community. By contrast, oil and gas development drags along with it the necessity of standardized rules and centralized control structures. Things have to run smoothly not just *here* in this place but across a network. There is a limit to which you can localize control over a global commodity production system. (Municipal fracking ordinances could be seen as the glocalizing of fracking—taking a standard product and adapting it to particular circumstances.) With fracking, no longer are we talking just about the work of one's hands and one's community. The Barnett Shale would not even be a viable resource if it were not for the expertise and equipment provided by a system that far outstrips the boundaries and capacities of a deme. It's kind of a paradox. In a way, the gas from the Barnett is more local than the vegetables at the Denton Community Market. But far more so than the tomatoes, harvesting the gas requires a vast technological network and system of control.

As I watched cattle grazing out the window, I thought this explained why the oil and gas business is so alienating in comparison with farming and ranching. The dying cattleman Homer Bannon says in *Hud*, a 1963 western set in Texas: "What can I do with a bunch of oil wells? I can't ride out every day and prowl amongst 'em like I can my cattle. I can't breed 'em or tend 'em or rope 'em or chase 'em or nothing. I can't feel a smidgen of pride in 'em 'cause they ain't none of my doing." Even if Homer could have formed a deme to govern those oil wells, he would still largely be an unqualified onlooker to a machine run by outside specialists.

This alienation is built into the technology itself. We can't really democratize it—it *demands* a technocracy, even a plutocracy. Whatever claims we might make against the system can be neutralized by saying, "Fine, but that's no way to run an oil and gas business." To commit to this form of energy harnessed with these technologies is to necessarily commit to a certain kind of political life that requires standardization and top-down control.[20] As Friedrich Engels once wrote, the machinery of the factory is more despotic than the capitalists. Even if we kick the capitalists out, we will still have "the authority of the steam" as our dictator.[21]

But maybe, I realized, things are not as dourly deterministic as all that. In the case of fracking in Pennsylvania and New York there are approximations to demes. They come into being when landowners gather together and form coalitions to gain collective bargaining power with the industry. These coalitions afford landowners a stronger voice in dictating the terms under which fracking will occur. One such neighborhood group, the Oakmont Energy Coalition, existed in Denton prior to my arrival in town and was so successful in controlling the terms of business that XTO Energy eventually decided they couldn't meet all of the landowners' demands for best practices and gave up trying to get leases signed. Of course, compulsory integration laws weaken the authority of these collectives, and membership in them is still dependent on the historical accident of owning mineral rights rather than the morally relevant criterion of vulnerability to harm.

Matt chimed in again, "That's sixty." We were getting close to Denton's municipal airport.

Demes are not as absurd as Socrates's proposal for rule by philosophers, but they are still impractical. The industry would hate them. The patchwork of regulations introduced by municipal ordinances is already a headache. Imagine doing business if every single gas well were subjected to the "ordinances" of a temporary political body comprised of those in close proximity. It would be wildly unpredictable. Hobbes would be rolling in his grave.

Then again, the trouble with laws is their rigidity. They can't be tailored to specific circumstances. For Socrates, a wise man's judgment would always be preferable to the iron cage of bureaucratic rules. Maile and Matt would argue that the judgment of the affected people is preferable. I understand that allowing their judgment to rule would be impractical, but aren't there higher values than practicality?

WE DROVE BY an old gas station that still had its Christmas message, "Happy Birthday, Jesus," on a little plastic sign advertising beer. I wondered how the rural libertarians living in the outskirts of the city would vote in November. And what would they think about demes? Maybe they'd prefer the invisible hand, supposing that the market is the only thing wise enough to sort out the complexities introduced by the unlimited pursuit of wealth and comfort in our high-tech society.

The leading libertarian philosopher Robert Nozick dismisses the idea that "people have a right to a say in the decisions that importantly affect their lives." He imagines an example where a woman is courted by four men. She will choose one to marry, and this choice will have a profound impact on all the men's lives. But "she has a right to decide what to do, and there is no right the other four have to a say in the decisions which importantly affect their lives that is being ignored here." [22]

Nozick is saying that every decision has "external costs" that are borne by a third party who does not have a say in the decision. There is no such thing as a purely personal choice. Libertarians seem to imagine a nightmarish scenario where we attempt to rectify this situation through a massive government that insinuates itself into every tiny decision in order to make sure all of the interests involved are represented. It would be Big Brother on steroids.

Yet, as I see it, the reality is closer to an opposite nightmare scenario where the powerful make decisions that impact the weak, but the weak have no political recourse to influence those decisions. At least the suitors had the institution of marriage and courtship to shape their interactions. Their relations with the woman were structured in a way that gave them a voice. Out in the balance, however, there is no "they" there to have a voice, because there is no institution to recognize the people as citizens.

"That's ninety," Matt reported. We could now see UNT's three wind turbines emerge on the horizon.

In a libertarian utopia, state universities like UNT would not exist. They want a "night-watchman state" or "minimal state," where government is limited to the only function it can legitimately serve, namely, protecting individual rights. But Sharon, Cathy, Maile, Matt, Calvin, and all the others impacted by fracking have a right to the peaceable and safe enjoyment of their property. This is a bedrock libertarian position. Nozick writes that the minimal state "treats us as inviolate individuals, who may not be used in certain ways by others as means or tools or resources; it treats us as persons having individual rights with the dignity this constitutes." [23] That's the basic principle behind informed consent as the first condition for ethical innovation. It may be that in an age of high technology, even the minimal state has to be big.

"That's 115," Matt said. In fifteen miles, he spotted 115 gas wells. We rolled down Jim Christal Road, which turns into Hickory Street as it heads to the courthouse on the downtown square. One of the last wells in Matt's count was Smith-Yorlum #7H, operated by EagleRidge Energy. Almost exactly a year earlier (in April 2013) that well vomited its contents—all 6 million gallons of frack fluids—high into the air for over fourteen hours in a spectacular blowout, or as one industry representative described it to me, "a fountain of water."

Part III

SEE NO EVIL

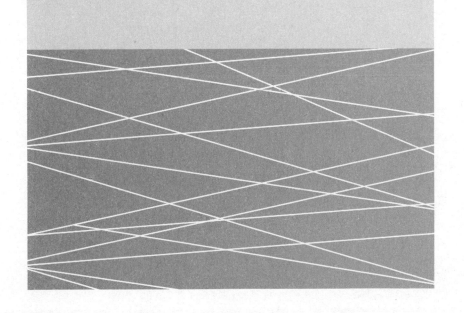

CHAPTER 7

HAND IN THE COOKIE JAR

WHEN THE VIDEO CAMERA TURNS ON, THE FIRST thing we see is Pam Brewer's back door. Things are shaky and slightly at an angle as Pam, who is holding the camera, leans forward with her free hand to grab the doorknob. Then we see an April morning in Texas. The horizon is turning orange, but the grass and oaks still appear more black than green. The swing set in her back yard is just now visible in the dim glow. A few hundred feet away, what appears to be a workover rig (a smaller rig often used to fix a problem with a gas well) is bathed in bright white lights.

Pam says in an exhausted voice, "This is what I've had to listen to all night long." [1]

A *CHAWUNCHA-WAWUUUUNCHA* noise is coming from the frack site.

It sounds like an industrial demigod is digging through a pile of stones in a metallic cavern. A plume of gray smoke is belching from large containers, and a geyser of what appears to be steam is spewing from the rig over a hundred feet into the air.

"All night long! This is psychological torture!" The volume of Pam's voice rises to match the devilish bellows. "I feel like I'm at the Branch Davidian," referring to the 1993 government raid of a fundamentalist religious compound in Waco, just a hundred miles south of her home in Denton. Indeed, the scene of the retching rig looks like something out of the Bible: "And if you defile the land, it will vomit you out as it vomited out the nations that were before you." [2]

About five hours earlier, at 1 a.m., the contractors had just finished hydraulically fracturing the Smith-Yorlum #7H gas well. As they were lifting out the pipe, a segment broke off (perhaps the threads were faulty or overtorqued), allowing frack fluids to rise uncontrolled under the pressure resulting from having just been squeezed into millions of minuscule cracks in solid black rock a mile below. The Barnett Shale has a reservoir pressure of about 3,500 psi. That makes it like a giant scuba tank full of a variety of gases, liquids, and sand with a very long straw attached. When they lost control of the well, all the volatile gases like ethane and methane pushed the fluids and chemicals back up the straw with enormous force.

Smith-Yorlum #7H is one of several dozen older wells in town that Mark Grawe at EagleRidge Energy had bought from the original operators. He wanted to convert them all from vertical to horizontal wells, basically taking the original straight straws and turning them into bendy straws with pipes stretching a half mile or more horizontally over a mile below the surface through the shale, that rich Oreo filling between the surrounding rock layers. Of course, when you are working in such extreme conditions, opportunities abound for "normal accidents." [3]

A few months before the blowout, I had visited another well just on the other side of Pam's house during drilling. In the trailer on-site, Mark showed me a computer printout that traced how the drilling pipe had behaved underground. They wanted it to make a clean right turn, but instead it took a left, looping up and around before heading in the right direction. The trace it left looked like one of those

Christian fish symbols you see on so many Texas pickup trucks. Mark shook his head and said, "Sometimes you just have to go where Mother Nature takes you." As Francis Bacon argued, "Nature, to be commanded, must be obeyed." [4]

Oil and gas operators in Texas are required to report the chemicals they pump into the ground at frack sites on the disclosure registry FracFocus (fracfocus.org). The entry for Smith-Yorlum #7H shows a total base water volume of over 6.1 million gallons. That includes 4,100 gallons of hydrochloric acid. And it includes over 5,700 gallons of proprietary and undisclosed ingredients, because operators can choose what they divulge to the public. In 2012, the year after the disclosure law was passed, companies filed for exemptions based on trade secrets 19,000 times. In the words of State Representative Lon Burnam, that's a loophole "big enough to drive a Mack truck through." [5]

That April morning in 2013, Pam didn't know this was a blowout. She thought it was just more of the aggravating business that is fracking. After all, she had been putting up with the commotion for weeks. So she put down her camera and got ready for work, kissing her daughter and newborn granddaughter good-bye. As she drove by the pad site, she saw the roughnecks sitting in their pickup trucks a good distance away from the rig.

Meanwhile, on-site, a remediation crew scrambled to solve the problem. By 8 a.m. Mark determined that the situation was uncontrollable and called 911. Emergency crews came racing down Jim Christal Road around 9:30. Firefighters banged on Pam's door and the doors of five nearby homes, ordering an immediate evacuation. "They told my daughter not to turn on the lights or anything, just grab her purse and go," Pam told me a couple of weeks after the blowout, "because any spark could have set off an explosion." Pam's daughter was so frazzled that she forgot to buckle the baby's carrier into the car seat. The fire department also established a no-fly zone around the area, which is close to Denton's municipal airport.

Remarkably, this happened just two days after a deadly explosion at a fertilizer plant in the town of West, Texas, a two-hour drive away. The blowout in Denton was nowhere near the scale of that accident, but it seemed to indicate the same underlying problem: a laissez-faire approach to development and lax regulations culminating in a thoughtless juxtaposition of industrial and residential areas. It was like slow-motion sleepwalking with a ticking time bomb.

When I talked to Pam, she said they had moved into that house ten years ago. "It was a quiet slice of Texas heaven in the open plains under the bright stars. My husband and I were going to retire here. But now with all the gas wells . . ." She trailed off with a sigh and then added, "We don't drink the water from our well anymore. I wonder: Am I being poisoned?"

Then Pam said something that made me think about Matt's vulnerability principle. "Suddenly I've got a gas well by my back door and another one by my front door. And no one asked me how I felt about it."

Apparently, no one cared how state regulators felt about the blowout.[6] The Texas Commission on Environmental Quality (TCEQ), which is responsible for regulating air emissions from fracking, was never officially notified of it. Darren Groth, head of Denton's Gas Well Inspections Division, told me that he got a call from TCEQ that afternoon. "They said, 'Hey, we're watching the news and we see this blowout. What's going on?'" Darren told them to hurry to the scene. They didn't take an air sample until 3:21 p.m.[7]

The well was capped eighteen minutes later at 3:39 p.m., after spewing thousands of gallons of fracking chemicals for over fourteen hours. No one really knows what came out of the belly of Smith-Yorlum #7H. Working with his own contractor, Mark guessed that 1,281 pounds of natural gas volatile organic compounds (VOCs) were released, and TCEQ accepted his estimate. When the fracking chemicals hit the heat and pressure of subterranean depths, they transmogrify into new compounds, but we don't know which ones and what they might do to people.

Nonetheless, on the basis of their assessment consisting basically of one eighteen-minute air sample, the TCEQ incident report concluded "no compounds exceeded the Air Monitoring Comparison Values (ACMVs) for long term or short term exposures."[8] TCEQ uses ACMVs as benchmarks for flagging problems and prioritizing the agency's limited resources. Though these numbers appear objective, they are full of judgments (e.g., what's acceptable: one case of cancer in ten thousand people or one in a million?). Despite these ambiguities, the report about the blowout confidently concluded: "No violations associated to this investigation." Buried on page 22 of the report, though, is the admission that "there were no monitors present during the event, and there is no monitoring data available."

So much of what passes as monitoring on the Barnett Shale is just this kind of false assurance where skimpy data are dressed up in official reports and adorned with impenetrable jargon, so it looks like the experts actually counted all the molecules to conclude that everything is all right.[9] As Rachel Carson put it, when the public sees a problem, "it is fed little tranquilizing pills of half truth."[10]

This brings us back to our analysis of real-world experiments. I've already argued that fracking often fails the first condition of ethical innovation—informed consent. In the next two chapters, I'll argue that it also fails the second condition, namely monitoring, or, as our friend Max More puts it, "inspecting the flowers for signs of infestation." I said earlier that the proactionary principle is kind of like a judge treating a technology as innocent until proven guilty, but it might be more accurate to say that the technology is like a teenager allowed to roam around on its own under close surveillance. If it's caught causing trouble, then steps are taken to reform its behavior.

We might also call the proactionary approach to innovation a sentinel system, from the way canaries were long used in coal mines to warn miners of the presence of toxic gases.[11] Rachel Carson most famously used songbirds as sentinels of a looming tragedy brought on by DDT. If we can't *prevent* the harms in advance, we nonetheless proceed cautiously, watching our canaries with great care, ready to turn back the moment they show signs of trouble. Of course, this shifts the burden of proof in favor of the powerful interests presently benefiting from the status quo. That's the problem, because our current innovation system is structured to make it exceedingly difficult to establish evidence of harm. It's one thing to presume the technology is innocent until proven guilty. It's another thing entirely to turn a blind eye.

Texas Sharon once told me that monitoring environmental pollution is like conducting mammograms: it's better to prevent exposure in the first place than detect harm after the fact. I share her sentiment. Still, there's only so much prevention that can be done. At some point, we have to gamble on a new technology. The question is: Can we keep tabs on the genie once he's out of the bottle?[12]

NOW FAST-FORWARD nine months to January 3, 2014. I'm standing with Maile Bush and Calvin Tillman on the corner of Vintage and Shagbark in the Meadows neighborhood. We're in a grassy median in the middle of the road about 30 feet from the south gas well site. The drilling is done, and EagleRidge is going to begin hydraulic fracturing in four days. In three days, DAG will decide to pursue the ban.

The wind is so slight that it barely moves Maile's curly hair as she speaks rapidly and nervously about her plans to evacuate their home during the impending hydraulic fracturing (I knew of two other families with the same plan). It's cold, but I never complain about cold in Texas—I try to absorb it as if I could store it in my bones and call it out into my blood when July comes. Yet part of me does wish I was wearing MG's new leopard print winter hat he got for Christmas, a hat

Family picture at the Don Deal #1 well near the right bank of the
Atchafalaya River in Louisiana in 1982, thirty years after my great-grandparents
struck oil there (and the same year that George P. Mitchell tapped into
the Barnett Shale in Texas). From left: My great aunt Alice
(Don Deal's daughter), my uncle Pete, my aunt Judy, my dad Bob,
and the Deal family attorney Bill Brinkhaus. *Courtesy Pete Boynton.*

Amber snapped this picture of me on one of our family fracking trips.
This is the site of the C.W. Slay #1 gas well, the discovery well for the Barnett
Shale, which was first connected to a pipeline in 1982. *Courtesy Amber Briggle.*

A typical scene on the Barnett shale: an old gas well site with rusty condensate tank and separator. I took this picture on a fracking tour with a former student, Matt Story.

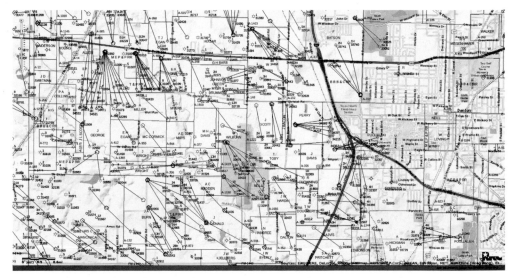

A screenshot of Denton from the Texas Railroad Commission's online GIS viewer. The pentagons are surface locations of horizontal wells; the lines radiating off the pentagons indicate the wellbore some 7,000 feet below the surface. The asterisks are the bottom hole locations for gas wells. An asterisk by itself indicates a vertical well. The circles are wells that are permitted but not yet drilled. *Courtesy Texas Railroad Commission.*

Does This
Mean Green?

www.dentondag.org

An Internet meme Amber and I created after my visit to the frack site next to the UNT football stadium in October 2013.

Drilling at the southern gas well site at the Meadows at Hickory Creek in the winter of 2013. Two frack sites went active in this residential neighborhood less than 200 feet from homes. This was after we passed an ordinance requiring a 1,200 foot setback distance. No one in the neighborhood was adequately notified that this could happen. This is when I began to think seriously about the ban—without it, this scene could be repeated across dozens of sites in the western part of the city. *Courtesy Gena Felker.*

Cathy McMullen gives a fracking presentation to a packed house in Denton at the Denia Recreation Center in January 2014, six weeks before we unveiled the Frack Free Denton campaign.

Cathy puts the very first signature on the petition to ban hydraulic fracturing in Denton, while Sharon Wilson captures the moment. This was at our petition launch party on February 20, 2014.

MG helped me gather signatures on the petition
throughout the spring of 2014.

Frack Free Denton marched in the Fourth of July parade around
the downtown square, handing out candy and literature.
Courtesy Roddy Wolper.

More than 600 people attended the July 15 hearing when City Council sent the ban to the general election. From left: Sharon, Cathy, me, Ed, Carol, Rhonda, and Sandy Swan. *Courtesy Amber Briggle.*

In September, three generations posed for the ban in our front yard. From left: me, Amber, Mema (Amber's mom, Barb), Oma (my mom, Rae), and of course MG and Lulu in the little red wagon. *Courtesy Bob Briggle.*

This is an ad we ran in October in the *Denton Record Chronicle* after I used the industry's own numbers to show how fracking contributed just $20 (out of $7,700 total) per student in Denton's schools. This was in response to the opposition's constant refrain that fracking brought major benefits to the local schools and economy. *Courtesy Frack Free Denton.*

Denton's Day of the Dead festival features a fun soap-box derby "coffin race" down Hickory Street. Jeff McClung built Frack Free Denton's racer, while Lulu and MG helped paint it.

Proud volunteers at the end of a long day on November 4 working at the polls.
From left: Matthew Long, Tara Linn Hunter, Michael Briggs, and Dani Parillo.
Courtesy Michael Leza.

My impromptu victory speech on the night of November 4 on the stage
at Dan's Silverleaf. Fittingly, the word "for" (as in vote for the ban) is on
my face from the television projector. Behind me, Cathy hugs Rhonda,
Sharon wipes away a tear of joy and relief, while Tara Linn and Ed
try to comprehend it all. *Courtesy Michael Leza.*

that put an end to his superhero phase and began a yearlong obsession with wild cats.

Calvin is cradling the pudgy belly of a summa air canister in his arms as if it were a little child. The summa canister looks like a small propane tank for a backyard grill. He was going to take a baseline air sample from the area, which could be compared with later samples taken during hydraulic fracturing. He held the canister away from his body as we waited for a Bridgeport Tank Truck to pass by: it was hauling wastewater from other fracking sites to an injection well a few miles west. When it was quiet again he turned the valve on top of the canister. It opened with a hiss, pulling air into the vacuum of its metal lung.

Calvin was raised in the tiny town of Jennings, Oklahoma. He literally grew up in the oil fields, tagging along with his dad, the area's best welder, when he went to fix broken pumpjacks. In 1990, at the age of seventeen, Calvin joined the US Air Force, serving for ten years until he started civilian life as an aircraft mechanic. He still works in aviation, and he and his wife, Tiffiney, are both active in their church.

In 2000, they bought some land in the Texas prairie. Unbeknownst to them, they had actually just moved into "downtown" Clark (later to be renamed Dish). This town, really just a collection of small homes on large lots, had incorporated hastily to ward off the voracious appetite of nearby Fort Worth, which was annexing surrounding properties. This pulled Calvin into local politics, which was soon overwhelmed by the booming Barnett.

As production soared, the industry determined that Clark was the ideal spot for a pipeline hub to transport the gas. Shortly after Calvin moved there, pipeline companies, deputized with the power of eminent domain, began building compressor stations around the area, often against the will of the landowners. The compressors were noisy, noxious, and prone to blowouts of their own. Dish had quickly gone from horse country to industrial zone. Calvin knew there were gas wells in the area when they moved, but he had that image in his head of the days spent helping his dad fix pumpjacks. The scale and intensity of the fracking in Dish were like nothing he had ever experienced. "This wasn't your daddy's drilling," he once told me.

Calvin, who became mayor of Dish in 2007, wrote a resolution urging the Texas State Legislature to give cities more control over pipelines. The resolution went nowhere. Then Dish released an air quality test conducted by Wolf Eagle Environmental that showed elevated levels of several harmful chemicals.[13] As a result, Calvin got TCEQ to install a permanent air monitor in town. Sensing a potential enemy, the industry tried to subtly court Calvin. They couldn't bribe a public official, but they'd tell him about how he was going places in the world and how they'd like to watch him succeed.

After we got our air sample, Maile walked back to her home, while Calvin and I headed over to his truck. We started talking about our kids. The Tillmans' first son, Clay, was born in Wisconsin ten weeks premature. "We'd visit him in the hospital," Calvin said, "and Tiffiney and I were heartbroken to see all the abandoned babies. We adopted Joshua shortly after we moved to Texas." It was the boys' recurring nosebleeds and Clay's worsening asthma that forced the Tillmans to move to Aubrey (east of Denton and east of the Barnett Shale) in March 2011, even though Calvin was still mayor of Dish at the time. One of his last acts as mayor was to install solar panels on their little town hall building. "I have zero doubt that those compressors and gas wells caused my boys' health problems," Calvin said as we crossed the road. "As soon as we moved, the nosebleeds disappeared and the asthma was under control."

"Ah, but did you have proof?" I said, mocking the industry's constant refrain about how there is no solid evidence that fracking is harmful.

"Proof!?" Calvin said, smiling back at me. "Who can afford it?"

In Aubrey, the Tillmans had a girl and named her Evie; she is just a couple of months younger than Lulu. I knew about the industry's attempts to win him over when he was mayor, so I asked, "Wasn't it tempting? You could have made lots of cash to raise your family."

Calvin looked up, and I could see the fog of his breath against the beige sound barrier around the frack site behind him. "Money's great. But I'd have to look my boys and my little girl in the eye every day and live with that choice. No thanks, man. Some things can't be bought."

That's when I first saw beyond the image I had of Calvin, the star of the *Gasland* documentaries. I noticed that he's actually a quiet man who carries a tender sadness in his eyes. He had become something of a celebrity, but he was really just a dad like me, patting his kids on the head, letting them sit on his lap during another meeting at City Hall, worrying about their future. What do we owe our children? We might just need faith that future generations can transform whatever mess we leave them into new opportunities. But my take on fatherhood is darker: our children will pay for the sins of their fathers. I feel a double burden, not only to raise my kids but also to help save the world for them. There was something about the way Calvin's shoulders rolled forward—not to mention the way he ran himself ragged taking on all this air monitoring as an unpaid second job—that made me think he felt the same weight.

Calvin was wearing his blue ShaleTest shirt and hat. ShaleTest is a nonprofit organization that provides free monitoring services for communities negatively impacted by fracking.[14] The idea is to help people get the evidence they need to make claims for compensation for nuisances and harms caused by fracking. Fed up with the lack of monitoring provided by TCEQ, Calvin founded ShaleTest in 2010 with Tim Ruggiero, whose Wise County home had been devalued 75 percent by nearby gas operations.[15] In fact, Tim didn't live far from Cathy's old house. The minerals under their properties were part of a huge swath owned by the same rancher, Herb Wright. Once Mr. Wright leased the minerals, Tim was basically powerless to stop Aruba Petroleum from fracking on his land. Indeed, he found out about their plans only after they had cut through his fence, letting his horses out in the process, and started surveying his property. (Tim noted wryly later on that he could have been justified in shooting them under Texas's "Stand Your Ground" law that allows for the use of deadly force to defend private property).[16] After two wells were developed on his land, chemicals linked to fracking appeared in his water.[17] Like Cathy and Sharon, Tim became a refugee of the shale. He relocated his family east of the Barnett to Pilot Point, close to the Tillmans' new home in Aubrey. His experience with fracking was so atrocious that he told me, "If you don't have mineral rights, you don't have any rights."

By the time I got to know him, Tim couldn't share the details of his story anymore, because he had signed a nondisclosure agreement, offering his silence in return for a settlement that helped pay for his family's move. Once when I asked him about this, Tim, who has a stocky build and intense eyes, looked at me square on and said, "Here's my quote: 'The matter has been resolved.' Period." Tim, his wife, Christine, and their daughter, Reilly, had become just one of hundreds of nondisclosure cases across the country. It's an industry strategy that keeps data from scientists, the media, regulators, and policymakers and that makes it all the more difficult to challenge their claims of fracking's innocence when it comes to air and water contamination.[18]

When we got back to Calvin's truck, I asked him about the early days of ShaleTest. "One of the first things we did," he said as he opened the door, "was tour the Marcellus Shale, telling Pennsylvania families about what to expect as fracking moved up there. We got to know Stephanie Hallowich and signed her on as part of ShaleTest." The Hallowich home in Pennsylvania had been surrounded by gas wells and pipelines. The family developed headaches, nosebleeds, and other symptoms. Like Tim and Calvin, Stephanie and her husband, Chris, complained loudly about the impacts to their air and water—until one day Stephanie called Tim and Calvin to say, "I can't tell you why, but I need to quit ShaleTest." Calvin said, "I knew right away she had signed a nondisclosure agreement."

Sure enough, the Hallowiches had been offered $750,000 by three fracking companies in exchange for their silence. Fearing for the health of their elementary-school-aged children, they took the deal. Later revelations from that court case showed that the industry's lawyer insisted that the gag order apply to the Hallowich children as well. They too could never speak about what happened to them. Court transcripts record Chris and Stephanie grappling with the implications of surrendering their children's First Amendment rights.[19]

In the warmth of his truck, Calvin labeled the summa canister and stored it in a box. Later that day, he would ship it to a lab in California that would assay the air inside for a suite of chemicals. Calvin grimaced as he packed away the canister. "Boy, I hope we got a good sample. It didn't seem to make the right noise when I opened it up." We wouldn't know the results for a couple of weeks, and by then the hydraulic fracturing would have started. He didn't have another canister with him, and besides, it would cost over $300 to use it. The air testing was being funded by donations from Maile and her neighbors, and they were running short on cash. They needed to save their remaining canisters for the fracking that was about to come.

A FEW MONTHS LATER, on a sunny Saturday in April 2014 and with the Frack Free Denton campaign off and running, Calvin let me tag along for another one of his ShaleTest trips. He was working on Project Playground, which entailed conducting air quality tests at several playgrounds near fracking sites.

One of the sites included in Project Playground is the purple park, so I rode my bike there to meet Calvin. When I saw him, he was fiddling with a big black suitcase in the backseat of his truck. Inside the suitcase was a forward-looking infrared (FLIR) camera designed to detect dozens of chemicals. He handed the camera to me and started explaining its features. The camera was heavy and cumbersome; I needed both hands to hold it steady as I looked through the lens. "Careful," Calvin said, "that's a $30,000 piece of equipment."

"What!" I said in a panic as I set it back in its case.

"Oh, yeah," he replied, patting the camera with a mechanic's knowing touch, "but that's nothing. This is a pretty basic model. They've got some that cost over $80,000."

Calvin hooked the camera up to a portable recording device in case he saw something he wanted to record. He took the camera and handed the recorder to me. "Hold this," he said and started walking. I tagged along behind him, holding the device tethered by a cable to his camera. Calvin stopped at a streetlight and hiked up onto its concrete base about three feet above the ground so he could get a better view above the stone wall surrounding the gas well site on the other side of the street.

"I was out here last Saturday and one of the condensate tanks had a leaking thief hatch. There was a pretty good plume of emissions coming out," he said down to me over the noise of the cars cruising along Bonnie Brae. Then he turned the bill of his cap backward to get his eye onto the camera lens, and there was a long silence as he scanned for signs of leaks. Not having technologically enhanced vision, all I could do was watch the cars go by. Everyone was gawking at us. One lady with big blond Texas hair scowled and honked the horn of her huge Ford F-350 truck. She obviously disapproved, but I doubt she had a clue about what we were doing.

"I always get strange looks with this thing," Calvin said as he climbed down from the light pole. "I tell people, 'Don't mind me, I'm just doing the job that TCEQ should be doing.' I mean, often when TCEQ investigates a complaint or conducts a study they give the operator advanced notice that they are coming. The company will just shut down until the sheriff leaves town."[20] And despite City Council's promise a year earlier to implement its own air monitoring program, Denton's gas well inspection team still did not have any monitoring equipment.

Calvin started walking back to his truck and powered down the camera, saying, "They must have closed that hatch. I don't see anything today." There was another site he wanted to visit down in Fort Worth, so I chained my bike to the light pole and joined him for the thirty-minute drive.

The Fort Worth site was massive. Behind the chain-link fence topped with barbed wire were four compressors, several lift compressors, twelve wellheads, and fifteen condensate tanks. At first, Calvin just drove slowly along the fence line through a scrabble of rocks and weeds while I looked through the FLIR camera out the open passenger window at the compressors. By then, I was less skittish about how expensive the camera was and more intrigued about how it could enhance my vision, making visible a world inaccessible to my naked eyes. After a while of getting adjusted, I suddenly did a double take. Without the camera, all I could see was blue sky above the compressors, but with the camera I could see thick, fast-roiling white flares, like the ghosts of flames. "That's probably just heat," Calvin said after I described what I was seeing. "The chemical emissions have a different look. See if you can find a slower-moving gray cloud that lingers longer." I scanned the condensate tanks and located just what he was talking about. There was a leak in one of the pipes above a tank.

We kept driving a bit toward another tank that was set apart and, at 20 feet tall, was twice the size of the rest. Through the FLIR camera, I saw a gray plume coming out of its lid and wafting to the south. At that point, Calvin parked the truck and gave me a yellow handheld device designed to detect VOCs. We walked to the southern side of the site, leaping over a puddle of water along the way that was streaked with an oily rainbow. A few feet later, we both smelled something sour and acrid, kind of an ashen brine in the nose. The digital display on the VOC meter ticked up several notches. A few seconds later, the smell would disappear and the VOC readings would go down. Then, when the breeze picked up, the smell and high readings would return.

Calvin wanted to take an air sample, so we went back to the truck to get the summa canister ready. He had a tablet with an app that allowed him to enter the GPS coordinates of the sample site, record weather conditions, and then upload everything into a server on the cloud. This site was just a few hundred feet from a playground and basketball court that served the low income, mostly minority neighborhood across the street. Three young boys were playing basketball as Calvin prepared the canister. On this day, the basketball court was downwind of the site. Through the FLIR camera we could see those gray plumes drifting toward the boys. We hopped back over the suspect puddle, waited for the VOC meter to spike again, and grabbed a sample in the canister.

There's something quietly heroic about Calvin and Tim's work at ShaleTest. It's part of the environmental justice movement to empower the disenfranchised. Yet for all that's good about citizen science, it's also an indication that something is amiss. Here's a basic principle: those running the real-world experiment should pay independent scientists to monitor it. We shouldn't have to rely on the kindness of donors and the volunteer hours of moms and dads like Calvin. In the absence of a comprehensive monitoring program, it's all too easy for the industry to dismiss any complaints—even those bolstered by citizen science—as just anecdotes. They'll say that all we have are isolated stories. And often they'll be right, because there's not enough money or institutional capacity to connect the dots.

That's certainly true on the Barnett, where TCEQ doesn't have anything close to an adequate monitoring network.[21] Dr. Ireland, the industry representative who served on Denton's task force, is fond of saying that the Barnett Shale is the most monitored air in the country. Then he cites industry-funded studies with millions of data points showing no pollution problems.[22] But what he doesn't highlight is that all those data points are from just *seven* monitors. That's seven monitors for over 17,000 gas wells and related facilities spread across the 5,000-square-mile Barnett Shale region. There is no way those monitors are picking up the chemicals wafting over the purple park in Denton or the basketball court in Fort Worth. Indeed, when it has conducted short-term, intensive studies using state-of-the-art mobile laboratories, TCEQ has found problems on the Barnett, including "some of the highest concentrations of benzene" in the state.[23] Some sites had sustained benzene concentrations of 93 ppb (parts per billion), which is sixty-six times higher than what TCEQ considers a safe long-term exposure level.[24] Other studies confirm the presence of high levels of toxic chemicals near frack sites.[25]

Calvin was trying to make sure that the general lack of evidence of harm would not get passed off as evidence of lack of harm. That is a Sisyphean task. It's not just inadequate monitoring and nondisclosure agreements. It's also the frequent lack of baseline data to help establish causality, the high costs of testing, the fact that gas well sites do not have to submit emissions inventories to state regulators, the trade-secret chemicals, the general unwillingness of state agencies (heavily influenced by the oil and gas industry) to study the impacts of fracking, and exemptions from federal laws.[26] All of these practices suggest what appears to be a deliberate unwillingness to look for problems. And they would all have to change in order to make a proactionary approach to innovation ethical.

ON MAY 6, a few weeks after my tour with Calvin, Denton City Council passed a temporary moratorium on new drilling permits and began another revision of the drilling and gas well production ordinance. By then, DAG was fully absorbed in the Frack Free Denton campaign and the task of gathering the requisite number of signatures on our petition. We had been so central to the earlier ordinance revision process, but were now on the outside. At that time, we were not interested in helping with revisions, because we had come to see the futility of that approach. As I explained in chapter 2, the cows were already out of the barn. More rules could only shut the barn door tighter, not corral the cattle back in. But we were concerned that voters might naïvely think that City Council would just fix the problems with a new ordinance, obviating the need for the ban. So we emphasized in our messaging (mostly our Facebook page, blog, and press interviews at that point) how the ban was our *last resort* after years of trying to write reasonable rules.

The next day, I rode my bike to City Hall to mark a major milestone in the campaign. I arrived a few minutes late, and Ed, Carol, Sharon, Rhonda, and Cathy were already out front getting their pictures taken and doing interviews with several regional news crews. I could tell Cathy was getting anxious, because the moment she saw me she said, "Enough already, let's do this," and she picked up a big banker's box in her arms and started marching inside. The rest of us, wearing our Frack Free Denton T-shirts, fell into formation behind her. I didn't even have time to clip my bike helmet to my backpack. Cathy was all business. The cameramen scurried to get in front of us, and we got a little taste of the paparazzi as they walked backward to film us walking into City Hall.

For over three months, Cathy had been keeping that box in the fire safe by her bed—losing sleep over what might happen to the signed petitions sheets it contained. We now had roughly 2,000 signatures, and the time had come to turn them in.[27] Cathy brushed past the cameramen to set the box on top of Jennifer Walters's desk. As city secretary, Jennifer would have the job of officially certifying the signatures and reporting to City Council that we had submitted a valid petition. Cathy said (loud enough for the news crews to pick up the audio), "We're here to turn in the signed sheets petitioning our government to ban hydraulic fracturing in the city limits of Denton." Jennifer blushed at the cameras, said, "Okay, thank you very much," and flashed a sheepish smile before hauling the box into a back office. According to Denton's charter, the City Council would then have sixty days to either adopt the proposed ordinance (the ban on hydraulic fracturing) as law or put it on the November ballot to be decided by the voters.

I couldn't get a read on Sharon's mood, because she was busy tweeting the moment. Rhonda, Ed, and Carol were all smiles. But I could tell that Cathy was torn by the emotions of the day. She had become so protective of those petition sheets that she almost didn't want to hand them over. Yet she also couldn't wait to be rid of them, not only because she was so worried they would be stolen or destroyed but also because this marked a new phase of our campaign. As she slid the box across the desk, the campaign itself slid from gathering signatures to winning votes. The stakes got much bigger. We now had a chance not just to write law but also to make history by becoming the first city in Texas with active gas wells to ban hydraulic fracturing.

The emotional wrestling match going on inside of Cathy gave way to a big old grin. The rest of us mirrored her smile—no matter what else might happen, we had at least gotten this far, which was no small feat. We hugged each other and savored the moment, but only briefly, because quickly Cathy turned to the assembled cameras to deliver an impromptu speech, saying, "We know the industry is going to outspend us. But we have one thing they can never buy: a community of caring people who will do what it takes to look out for each other." I would have never guessed just how prophetic those words would turn out to be.

Three days later, I went to former City Council member and local historian Mike Cochran's beautiful house on Oak Street near the downtown square. It was the night of the City Council election, and Mike was hosting the watch party for mayoral candidate Chris Watts. Chris had asked for my endorsement shortly before we announced the ban. I was afraid that he might regret my involvement in his campaign after he learned of my intentions to push for the ban, because now he would be tethered to this "radical professor." But he never wavered, and although he never officially spoke in favor of the ban, he always said in public that he welcomed the democratic process of the petition and understood the frustrations that led to it.

Those were about the warmest words any council candidate had for the ban during the election season—although at one unusual candidate forum held at Dan's Silverleaf, a popular bar, Kevin Roden (who was moderating, because he was not up for reelection that year) got all the candidates to say that they would vote for the ban *as private citizens* when it came to their personal decision in the booth on November 4. Still, for the most part, the candidates were pretty cautious about the ban as a public policy. Indeed, Chris's opponent Jean Schaake said during another forum, down at Robson Ranch, that the ban would bankrupt the city and that we needed to take our concerns to state legislators in Austin. Considering the enormous influence of oil and gas money on the Texas legislature, I almost laughed out loud when she said this. There was no way the state was going to intervene to help Denton residents.

So, most Frack Free Denton supporters gravitated to Chris's camp. Indeed, Maile, Debbie Ingram, and several other Meadows residents were also at his watch party. As we sat around nervously sipping on margaritas awaiting the voting results, I introduced Maile and Debbie to Elma Walker. As they chatted, I couldn't help but wonder at the significance of their encounter. Here was a woman (Elma) who lived in Robson and worked with the city on its earliest fracking regulations

talking to Debbie and Maile, whose neighborhood triggered the call
for the ban many years later. And it was all at the election party for
Chris, who had represented the Robson area as City Council member
and now might lead the city as mayor through the process of deciding
on the ban.

When the early voting results were released, there was huge cheer
and an electric surge of energy at the party. Chris had carried the early
voters by a two-to-one margin, guaranteeing that he would win in a
landslide. I was chewing on a celery stick during his victory speech
when suddenly he pointed to me and thanked me for helping him win.
I almost choked. I was on the inside of power in a way I'd never been
before.

Around the time of the election, DAG issued a press release with
the results from one of Calvin's air samples taken during hydraulic
fracturing in the Meadows neighborhood.[28] The sample was taken
near the north site, not the south frack site where I had joined Calvin
back in January. It was literally taken in Debbie's back yard, next to
the fountains and garden she could no longer enjoy due to the nuisances of fracking. We also released some FLIR camera footage from
a compressor that EagleRidge had built on the south site.[29] That well,
#4H, struggled to produce—in its first month it yielded only about
half the amount of gas as that produced by #3H on the north site. So
it appeared that EagleRidge was trying to boost performance with a
compressor, which became a permanent noisy and smelly addition to
the neighborhood.

The data from the canister showed benzene in the air. And Sharon's FLIR camera footage showed a ghostlike vapor wafting off the
compressor against the pitch-black background, which is the way the
camera sees what looks to us like a blue sky. Cathy said the odors had
been worse at other times, but we didn't have the funds in place or the
qualified people to do the air tests when the best opportunities struck.

We also only had the camera every once in a while. It rotated around the country, spending some time in Texas with Sharon and Calvin before moving on to Wyoming, Pennsylvania, and elsewhere.

Though the media played up the results, EagleRidge calmly swatted them down.[30] Our figures only showed a slightly higher level compared to long-term exposure limits, but this, they said, was a short-term event. Besides, their press release continued, benzene can originate from many sources. It could have come, they said, from traffic, nylon carpeting, or paint remover. As for the FLIR video, EagleRidge wrote, "DAG [does] not explain how the video shows the emission of toxic chemicals."

Now, the video did not just show heat, but we didn't know exactly what chemicals it did show, because the camera is not capable of identifying the compounds it detects. When I told Jay Olaguer about this episode, his response was that the summa canister and even the FLIR camera were too low-tech. Jay is an air quality scientist with the Houston Advanced Research Center (HARC), which provides scientific assessments of energy, water, and sustainability issues. HARC was founded by the father of fracking, George P. Mitchell, which gives it credibility with the oil and gas industry. But it produces independent scientific assessments, some of which come to conclusions favorable to environmentalists. Jay had recently written one such study for HARC that drew heavy fire from the industry, because he concluded that fracking on the Barnett Shale was a significant driver of the region's ozone problem.[31]

Jay spends lots of his time driving around in a mobile laboratory in the notoriously polluted Houston Ship Channel. Whereas a summa canister only grabs what drifts by it, the equipment on the mobile lab can scan for emissions across an entire study area. It can also detect more chemicals and speciate (identify) them in a way that a FLIR camera cannot. The FLIR camera DAG used can only show the presence of chemicals. The mobile lab can tell you which ones and at what concentrations. I would later work on a grant proposal with Jay to try to bring his mobile lab up to Denton. Here is a sample of some of his text about the lab:

> An Ionicon Proton Transfer Reaction—actioncon Proton Transfer Rb can tell you which on$^+$, O_2^+, and NO^+ to be used as reagent ions in addition to the hydronium (H_3O^+) ion. H_3O^+ will be used to measure a suite of VOCs with proton affinities near or above that of water, including benzene, toluene, xylenes, propene, aldehydes, and carbon disulfide.

Now that's high tech. But the problem is that it's also high cost, making $30,000 for a FLIR camera look like chump change. And the mobile lab requires a model to attribute sources of emissions. Even if we could drive around the Meadows neighborhood with Jay's big white mobile lab, the industry could still question the model.

There is no certainty to be had. Detecting chemicals in the air is nothing like counting apples in a basket. There will always be technical limitations, modeling assumptions, and methodological choices to debate. What this means is that if you've got a system that treats a technology as innocent until proven guilty, then the jury is going to be out for a very long time.

ESTABLISHING EVIDENCE OF guilt in real-world experiments is made difficult by two problems: causation and money.[32]

In graduate school, I worked in a biology lab trying to figure out what was killing the boreal toad, an amazing species of amphibian that lives at 10,000 feet of elevation, somehow managing to mate and mature in the short growing season of the high country. Populations were crashing, but we didn't know why. In the lab, we had dozens of cages, each holding four toads, and we'd hold all the variables constant except for the one we wanted to test. So they would all have the same amount of food and water and live in the same temperature with the same light/dark cycle, etc., but for some of the cages and not the others (the controls), we'd tweak one of the variables. In this way, we could isolate cause and effect.

In the real world, things are more complex. The lab is like the museum in the way its isolation can amount to a fiction. Out in the ponds on the mountain it could be a complex combination of factors behind the demise of the boreal toad. Or it could be something we didn't think of. This is the problem of causation: when you can't control the variables, establishing cause and effect becomes a nightmare. Different studies will yield different conclusions that different interest groups can cherry-pick as evidence to support their side. This is precisely the case with fracking and air quality on the Barnett.[33] In this way, rather than clearing away political controversies, scientific research can actually make them worse.[34]

The problem of causation lurks underneath the history of industrial impacts on communities. One example is the small town of Fallon, Nevada, where from 1997 to 2002 sixteen cases of acute lymphoblastic leukemia were diagnosed in children. Health officials declared it the most significant childhood cancer cluster on record, but a major study was unable to determine the cause. Despite hopes for answers, the parents of sick and dead children were left with only "an enduring sense of uncertainty."[35]

Scientists suspected the cause might be jet fuel, leaked from a pipeline that carried 34 million gallons of it through Fallon annually. The fuel contained benzene, a carcinogen. Yet as scientists compared blood and urine samples between sick and healthy kids, trying to pin down the fuel as the culprit proved impossible. There were too many uncontrollable, confounding variables: genetic differences, lifetime exposure to other pollutants and pathogens that could affect the immune system in poorly understood ways, and behavioral patterns. On top of this, many of the chemical suspects had unknown toxicological effects. Tungsten showed up in blood samples, but even ten years *after* the study in Fallon, it was still sitting on a list waiting to be analyzed by the National Toxicology Program.

Science is slow, and it needs large sample sizes to come up with statistically significant findings. Cancer clusters are often so small that studies result in false negatives, finding no significant link between a toxin and a disease even when one might exist. By the time you can finally draw a definitive link, as one public health scientist said, "you're basically doing a body count." [36]

The second challenge to gathering evidence is money. Whether it is jet fuel in Fallon or fracking in Denton, there's an industry profiting from the status quo and interested in prolonging uncertainty about any evidence of harm. To put the point the other way, the industry will claim with certainty that what they are doing is safe. In the early days of DAG, I was invited to a stakeholder conference hosted by one of the region's top oil and gas companies. They were trying to figure out how they could improve their public image and their social license to operate. As I sat at a table with executives, I suggested they could use less toxic frack fluids. The room fell silent until a manager stammered in disbelief, "Are you suggesting there's something wrong with our fracking fluid?!"

If proaction is going to be a viable ethic for innovation, it must take place in a political context that shelters science from the depredations of corporate interests. The lack of this shelter has been the subject of several books that detail how corporations have clouded scientific evidence of harm by promoting unwarranted doubt, passing bogus studies off as scientific, and manipulating the media's "fairness doctrine" to gain credibility for unfounded claims.[37] Some, like Ed, call this the "tobacco strategy," because it was first used by cigarette companies to forestall regulations in the face of mounting evidence that their product was causing cancer.

To use frogs as an example again, consider the case of atrazine, an herbicide manufactured by Syngenta and applied to more than half the corn in the United States. In 2000, when the biologist Tyrone Hayes discovered that atrazine might impede the sexual development of frogs, Syngenta began a campaign to discredit his work.[38] They also produced their own studies to prove atrazine is harmless. By 2003, the EPA approved the continued use of atrazine due to remaining uncertainties about the links between it and frog deformities. That same year, the EU, taking a precautionary approach, chose to remove atrazine from the market.

The most troubling thing about the so-called tobacco strategy is not industry-funded junk science. Rather, it's the way the industry weaponizes scientific objectivity. In 2000, a consultant and lobbyist for Syngenta proposed language for what would become the Data Quality Act, which requires that regulatory decisions rely on research that meets high standards for "quality, objectivity, utility, and integrity." Corporations have since used the Data Quality Act to limit the impact of scientific results that conflict with their interests, because it allows them to claim that those results were derived from insufficiently rigorous and objective studies.[39]

Forget the nosebleeds of Maile's children or data from Calvin's canisters; even peer-reviewed science is often deemed too suspect to count as real proof. This part of the strategy relies on a philosophical point made famous by René Descartes: all knowledge that comes to us through our senses is open to doubt. We can even doubt whether the external world exists at all, let alone whether an herbicide is responsible for deformed frog gonads. Perhaps even more important than the question of who bears the burden of proof, then, is what the standard of proof should be. The industry would like to set the bar very high.

IN JUNE, after we had turned in the petitions, I met up with Calvin again at Royal's Bagels for a cup of coffee and the best cinnamon rolls in Texas. Calvin wanted to hear how things were going with our new "get out the vote" phase of the Frack Free Denton campaign. I told him, frankly, that we were struggling and feeling overwhelmed by the daunting task of mounting a massive ground assault. We knew we'd only be able to afford one or two mailers and a few ads in the paper. So we had to mobilize volunteers to distribute our literature . . . but how?

Calvin offered some helpful tips but mostly just lent me a sympathetic ear. Then I said, "You know, I'm writing a book about all of this." That intrigued him, and he started peppering me with questions and recommending books for me to read. When I tried to explain this part of the book, about monitoring and inspecting real-world experiments, he stopped me and said, "Oh, you've got to use the Parr case as an example of that!"

Most people think that Cathy's former neighbors Bob and Lisa Parr got a court to agree their health had been harmed by fracking near their home outside of Denton. But the jury ordered Aruba Petroleum to pay them $3 million to cover the costs of nuisance and trespass, *not* health impacts. Calvin put it this way: "In order to prove a company harmed your health, you have to get a clear picture of them with the lid off the cookie jar, their hand in the cookie jar, and then the cookie in their hand. Definitively connecting the dots like that takes lots of time and tons of money to cover the tests and legal fees. Hardly anyone is able to mount a case against those odds." [40]

I spoke a little while later with Bob Parr, who confirmed Calvin's remarks. "Shoot," he said from under his black cowboy hat, "it took three years for us to even bring that case to trial. We won the first round, but our lawyer told us not to hold our breath as it goes into the appeals courts. The justice system is greased with oil money around here!" He winked and rubbed his fingers together to symbolize the money for judges' campaign contributions. He seemed remarkably jovial and composed for a man likely to have a $3 million case overruled.

How high we set the bar is a matter of judgment. There is no objectively right answer for what standard of objectivity to use. We are doomed to philosophize.

The problem is that we forget all the philosophy—all the judgments—that gets built into numbers. As one senior official at the EPA noted, regulatory decision models can appear "objective and neutral," but in reality they "quietly condone a tremendous amount of risk." [41] But we usually don't see the risk, because we are blinded by the veil of expertise.

That's the point that University of Texas researcher Rachael Rawlins made in a hefty review of science and policy on the Barnett Shale that came out during our push for the ban. She showed how the Texas Department of State Health Services (DSHS) systematically and nontransparently employed conservative assumptions while investigating a potential cancer cluster related to fracking in Flower Mound just south of Denton. She concluded that DSHS was "most concerned with avoiding an error that mistakenly maligns the industry, rather than avoiding an error that mistakenly dismisses the concern that children are suffering increased rates of cancer." The hidden and biased judgments used by the state "created a false sense of security." [42]

All of these impediments to learning from real-world experiments highlight the bigger stakes for our fossil-fueled civilization and its climate-altering powers. Recall that "monitoring" shares the same root as "monster," meaning "to warn." As the powers of technology grow, they change the environment in ways that are often imperceptible to the unaided human senses. It's not immediately apparent that there is a problem, unless there is a network of instruments to detect the invisible changes and sound the warning.

In *Collapse*, Jared Diamond looks at what doomed past civilizations like the Anasazi and the Maya. [43] One factor that strikes me as salient for our civilization is a failure to act in light of warning signals. We often talk of the "body politic" as if society is like an organism with its political leaders as the brain. Monitoring acts like the eyes, sending signals about the outside world to the brain. If an organism, say, a rabbit, sees a big rock ahead, it will change course as it runs along.

The problem is that society isn't an organism with a unified sense of self and purpose; it is a collection of conflicting interest groups. Now, maybe a perfectly free market could spontaneously coordinate the movements of the body politic. Maybe it could make us as nimble as the rabbit, capable of avoiding the rock in our path. But that's not our reality. We've got very powerful groups profiting from our current trajectory, and they don't want us to see the rock.

CHAPTER 8

MAKING RAIN

DEVIN TAYLOR IS THE DA VINCI OF DENTON. Stocky and imposing, with a six-foot four-inch frame and a bushy beard, he can fix just about anything and has an uncanny, encyclopedic knowledge.

Two examples. In June, about a month after turning the petition to ban fracking in to City Council, the Briggle family volunteered on his family's farm near Era, northwest of Denton. Devin was going to get married out there in November, and there was lots of cleaning up to be done. The next day, I had horrible bug bites all over my legs. I saw Devin on the square for Twilight Tunes that evening, and while I scratched myself raw sitting on a picnic blanket he told me all about chiggers (you itch, I was delighted to learn, because the bug crawls in your skin and dies). A few days later, I rode my bike to a community Better Block meeting about what to do with the abandoned Piggly Wiggly off of Sherman Drive. Devin eyed up my red Schwinn for a moment and then launched into a history of Schwinn, a mechanical explanation of the different styles of gears, and a market analysis of the main bike brands. And don't even get him started on the history and mechanics of guns, the inner workings of computers or municipal utility districts, or basically anything else.

I always end our conversations with the same question: "How do you know all this stuff?" Each time, Devin, the gentle giant, gives the same sheepish grin, suddenly realizing he's been swept along the current of his own river of knowledge.

Every year, Denton holds an old-fashioned small-town Fourth of July parade with fire trucks, bands, horses, and Corvettes circling the courthouse square. Throngs of people line the streets, sweating and chasing their children diving here and there to catch the candy thrown by the folks in the parade. It's when I try, always unsuccessfully, to channel the cold I absorbed in the winter months.

I had reserved a spot in the parade for Frack Free Denton. Zac Trahan, with the nonprofit Texas Campaign for the Environment, let me borrow a wooden model of a drilling rig he had made. It broke down into three pieces that, when assembled, stood 15 feet high. My goal was to build a giant platform with wheels, put the rig on top of it, and pull it down the street in the parade. I had no idea how to accomplish this, but I knew exactly who to ask.

I sent Devin a message on Facebook asking how to assemble a rolling platform. Here is a snippet of his response:

> . . . wheel angle of attack is going to be important when dealing with rolling ease and stability. Rolling over a bump or obstacle is equivalent to rolling up a slope equal to the angle of attack for the obstacle. At an obstacle height of approximately 1/3rd of wheel radius the angle of attack is 45 degrees . . .

Devin told me what kind of wood to buy and sent me to Harbor Freight to get 10" caster wheels for the platform. He then took time away from rebuilding his house to help me assemble what he called "the giant skateboard," a 5-foot-square rolling platform. Of course, he had to supply the tools too, because, well, I'm a philosopher. Devin's fiancée, Alana, and her daughter, Zarian, came over to hang out with Amber, MG, and Lulu while we hammered, drilled, and sawed in the driveway. Except for the mosquitoes, we had a good time.

But Devin couldn't stay long enough to finish the project. He had to get home and prepare for an upcoming Planning and Zoning Commission meeting. By then, Amber had also joined the P&Z Commission, so every other Wednesday night she got to absorb Devin's freakish knowledge of plats, building codes, and floodplains. Amber had started bringing home a giant binder labeled "Confidential: Draft Gas Well Ordinance" from her P&Z meetings. It contained drafts of the ordinance that the city was once again revising even as we pursued the ban. Her position on P&Z and my position on Frack Free Denton made for some potentially charged encounters like this. I never touched that binder.

I tried to finish the giant skateboard on my own, but naturally I screwed it up. So I called on our neighbor Cal. Now, Cal is Texas. Divorced with grown kids, he lives alone across the street. He drives a pickup truck to his job as a precision engineer, and on the weekends he drives a Harley. He keeps only one brand of beer, Budweiser Select, in his fridge, and if you offer him the kinds of microbrews I like, he'll drink them but he'll give you a hard time about being too fancy. Cal collects and refurbishes guns, and over the years he has amassed an arsenal. If Denton is ever attacked, we should assemble at Cal's house to take our last stand. He has guns in a giant safe, more in a closet, and a couple more semiautomatic rifles under the bed. His shelves groan under the weight of boxes of ammunition.

In short, Cal is everything I am not. But we get along great. I consider him a good friend. When we moved to the neighborhood, he instantly adopted us. He helped us unload our moving truck, and over the years he has fixed our garage door, assembled the kids' bunk bed, mounted our television, and so much more. I am constantly borrowing his chain saw and ladder.

When I asked Cal for his help with the giant skateboard, I told him what it was for. He took a puff on his cigarette and gave me his skeptical look. "I don't know, man, I might kind of sort of be in favor of fracking." For the first time, I was afraid that a political issue would divide us. We clearly voted differently, but that never mattered because we were neighbors. Who you supported for president is beside the point when you're helping to cut up a tree limb knocked down in a storm.

I didn't know what to say, but finally suggested, "Well, I completely understand if you'd rather not help out." But then Cal gave me a punch on the shoulder and said, "Let's get to work." As we finished up the giant skateboard, I told him all about fracking in Denton and why I thought we needed the ban. I don't think I convinced him, but he was nice enough to listen and to offer his perspective in return. After we were done, Cal pulled MG and Lulu around on the giant skateboard. I don't think there's ever been a finer neighbor.

I wanted a slot in the parade because City Council was going to hold a public hearing about the ban on July 15. That's when they would decide whether to adopt the ban as city policy or to send it to the November ballot. At Cathy's suggestion, I printed up five hundred postcards to hand out as we walked in the parade.

I met Ed and Sharon downtown on the morning of July 4 to assemble the rig, put big "prohibited" signs on it (a red circle with a red slash through it), and attached it to the oversized skateboard. We had one moment of panic when we had to scoot it around a pack of Harley riders (praying with all my might we didn't scratch their motorcycles) and ran into a tree limb. The rig leaned back and nearly toppled over onto Ed, but it just barely stayed in place thanks to Cal's idea to use hose clamps and L-brackets.

We made it to our staging area, right behind a band of tubas and accordions. I gave Lisa Parr and her daughter, Emma, the honor of holding the "Pass the Ban" banner to lead our contingent. About thirty people marched with us in the parade. Rhonda brought a huge bucket of candy that we set on the giant skateboard under the rig, and the kids who joined us would grab handfuls from there and toss them to the crowd. Ed put on a Revolutionary War suit and strapped a drum to his chest. He played a battle hymn as we marched. It put me in a solemn mood as I pulled the rig by the courthouse. We were sounding the drums of war and parading our army. I was apprehensive about how the crowd would receive us. To my delight, we got lots of cheers, and we handed out all five hundred cards before the midway point of the route. People encouraged us along, shouting things like "Yeah! Don't frack with our water!"

Throughout the signature-gathering portion of the campaign, media coverage and controversy had been pretty light, except for a few key moments like when we turned in the petition. But all of that was about to change. As we were marching, Barry Smitherman, the chair of the Texas Railroad Commission, was drafting a four-page letter addressed to Denton's City Council. He mailed it to them a few days after the parade. In the letter, he requested that City Council "NOT approve" the fracking ban, which he characterized as "extremely misguided." He went on to discuss the virtues of fracking for economic growth, job creation, school funding, energy prices, climate change, and US energy independence from Venezuela and the Middle East.

Smitherman claimed we were pursuing the ban "without citing any concrete examples of hydraulic fracturing negatively impacting public health." Of course, the testimony of Maile and other moms in the Meadows neighborhood didn't count as evidence for him. Neither did the air samples collected by Calvin. Like TCEQ, Smitherman's Railroad Commission was ostensibly in charge of finding evidence of harm. But there are nearly 2,700 gas wells per inspector, and each year roughly 130,000 gas wells go uninspected.[1] To use Max More's metaphor, after not looking for weeds, Smitherman concludes there are no weeds. Meanwhile, in a patent case of regulatory capture, Railroad Commissioners collect 80 percent of their campaign financing from the oil and gas industry.[2]

At the end of his letter, Smitherman pointed to allegations that Russia was "secretly working with environmental groups" and suggested to City Council that perhaps "out of state groups" might be behind the effort to ban fracking in Denton. In effect, he was accusing DAG of being funded by Vladimir Putin.

Kevin Roden blasted Smitherman with a public reply on his blog:

> When our community has experienced blowouts, spills, and significant air quality concerns stemming from the very industry you are supposed to be regulating, where have you been? You've been silent, absent. . . . Yet a group of concerned citizens . . . frustrated with an obvious lack of regulation, take matters into their own hands and only then do you enter the discussion—and only to advocate squashing their efforts. You offer no policy suggestions, no empathy, and no sign that you understand the situation we find ourselves in here in Denton.[3]

Kevin concluded by calling Smitherman an "out-of-touch Austin bureaucrat." The first shot in the war had been fired.

ABOUT A MONTH BEFORE the parade, the natural gas lobbying group Clean Resources had hosted a free dinner discussion at Cartwright's Ranch House on the square. It was another attempt to appear local and folksy—grab some chicken fried steak and listen to Dr. Ed Ireland (one of the industry representatives on the task force) sing the praises of fracking. It felt like an invasion of our home turf. So Amber and I brought MG and Lulu just to check it out.

The event was sparsely attended by a couple of families and a few crotchety old men. After folks had gone through the buffet line, Dr. Ireland stood up and made a presentation. Through a combination of games on smart phones and coloring books, the kids stayed remarkably quiet. Dr. Ireland's speech was quite reasonable. He just framed his presentation around all the positives of fracking and either downplayed or ignored the negatives. The impacts of fracking are open to interpretation—what is included and excluded and in which context are they placed? Again, this does not make them subjective; certainly, facts are not negotiable. But it does mean that even seemingly simple-sounding questions are not so simple after all.

When it came time for the Q&A, a young father who had brought his son said, "I sometimes see big hoses hooked up to our fire hydrants around town. They're bringing water to frack jobs. How much water does fracking use in town?"

Dr. Ireland rightly replied that compared to watering lawns, fracking consumes a minuscule percentage of Denton's municipal water supply. Of course, what he left out was the fact that the water used on lawns remains in the hydrologic cycle, whereas the water used for fracking in North Texas is effectively lost forever, because it is contaminated and dumped down a disposal well. Again, both perspectives have validity. Fracking is tiny in context, *and* it is a uniquely permanent kind of water consumption. To those who say that the policy should be based on facts, the appropriate reply is often: Which facts?

But then Dr. Ireland added a twist to his narrative that I had not heard before. "You have to remember that when it comes to water consumption," he said, "when we combust natural gas, which is CH_4, one of the products is water. So the more gas we consume, the more water we are actually creating." The profundity of this point seemed to elude the old men as they gnawed on their chicken fried steak. But it hit me hard. Like God, they are *creating* water. Fracking is the great rainmaker. It is a brilliant illustration of the proactionary mindset and its picture of humans as cocreators.

Some people tried to rebut this point. But I thought Dr. Ireland was right. They are in effect making new water when CH_4 combines with O_2. It wasn't until MG and I took a little road trip a few days later (in late June just before the parade) that I spotted the underside of this rainmaking mentality.

A recent article in the *DRC* discussed the possible link between fracking and groundwater depletion in Montague County, just northwest of Denton.[4] It featured a man by the name of Terry Fender who struck me as the Calvin Tillman of water. Whereas Calvin captured metal lungfuls of air in summa canisters, Terry used a sonar device in a blue box to send sound waves slowly pulsing down water wells to measure the hidden heights of subterranean rivers. After fracking started around his home, Terry's own water well dropped 70 feet (from 95 to 165 feet below the surface). Neighbors reported similar drops. One of them, Robert McPhee, said that Terry's measurements proved that fracking's industrial-powered extraction from the aquifers was "definitely responsible."

But the drought in those parts was so bad that Wichita Falls (in western Montague County) was being forced into drinking recycled toilet water.[5] They were in a "stage 5 drought catastrophe," which meant no outdoor watering, with fines up to $2,000. So how could they be so sure fracking was the culprit? After all, the industry would say this was the causal fallacy. Benzene in playgrounds can come from treated wood and nearby cars. So too, declining aquifers may correlate with fracking, but that doesn't mean fracking is the cause. As one physician said, it is "the result of almost irresistible errors in perception" to see coincidence as cause for alarm and blame.[6]

I called Terry, and he invited me to tag along on one of his trips. I wanted MG to learn something about water, so I convinced him to come along—with the help of a promise to get ice cream on the way home.

Terry asked to meet us at the little town of St. Jo, which is about halfway between Denton and Wichita Falls. On the way out there, I tried to get MG interested in water. It was tough to do, because clean, plentiful water is just an assumed fact of his life. I asked him to remember that time our bathtub faucet was broken and we had to fill the tub with buckets. "That's how lots of people live every day—walking to fetch water, wondering if they will have enough." "Yeah, but, Dad," MG replied as he built his elaborate digital home in Minecraft, "we just got the water from the kitchen sink." It's ironic. Water is so precious that we secure it and make it readily available, but that makes it impossible for us to appreciate how precious it is.

I regrouped. "Okay, but let me tell you about the Trinity Aquifer," I said. "It's a giant underground lake just waiting for someone to stick a straw in it and suck up the water." That piqued his curiosity, so I continued, "Now, our water in Denton mostly comes from Lake Ray Roberts and Lake Lewisville on the Elm Fork of the Trinity River. But the folks out here have their own private wells with pumps that bring up water from the aquifer." As we drove past Muenster, the hills wooded with blackjack and post oaks reminded me of the old farm Amber and I used to work at in Pennsylvania.

After a while of silence, I said, "Look at those clouds, like big balls of cotton." MG looked out the window. Eager to teach me something, he said, "Clouds are made of water." Feigning, I replied, "What!? How can that be? I thought water stayed on the ground." Excited now, MG said, "It's called the hydro . . . er, something cycle . . . " And we spent the rest of the trip wondering how water can move through, over, and on top of the earth. Now water had lost its mundane mask as the stuff out of the faucet. It was blood, milk, and ocean. It was all that was on the first day of creation until a vault separated the waters above from those below. And ever since then, the waters have rebelled, like separated lovers reaching up in evaporation and running back down in chaotic rains . . . rains that had been too long gone from this parched land.[7]

MG and I met Terry in St. Jo, right next to the old wooden water well in the center of the downtown square. A thin and jovial man with a blue baseball cap atop his close-cropped gray hair, Terry seemed thrilled to have some company. We hopped in his truck, and he said early on that he thought the rule of capture was at the root of the problem. "The fracking industry has pumps that pull out 150 gallons every minute, and before that water is even in their holding ponds it belongs to them." The man who retained all the mineral rights under Cathy McMullen's and Tim Ruggiero's land would allow fracking everywhere . . . except, that is, on the fifty acres he kept around his home. But on that land, he allowed the operators to pump out as much water as they wanted. Cathy guessed that he made over $10,000 every month just selling water. It was a potential tragedy of the commons where shortsighted decisions about private property (in this case, water) might lead to major public problems—the Trinity Aquifer seemed locked in a long, steady decline.[8]

As we bumped along little county roads through forests and fields, Terry told us the story of Roger and Glenda Swaim, who lived in an unincorporated area of western Montague County, "in the balance," as Matt Story would say. They started having problems with their water well after their neighbor allowed EOG Resources to engage in industrial-scale water extraction on their property. The water level got so low that they had to replace two pumps that had clogged up with mud. When their 325-foot well went dry, they drilled a hole to 500 feet, but it too was bone dry. Roger and Glenda were forced to buy a holding tank and have water trucked to their home. Eventually they moved to Bowie so that they could be on the municipal water supply.[9]

Terry retired in 2002 on sixty acres near his wife's old home in St.

Jo. Looking for some worthwhile activities to fill his time, he joined the Montague County Property Owners Association. When he and his neighbors started seeing their water levels drop, he bought an instrument and measured about sixty wells. Terry said, "I wanted to get something like a baseline sample to provide evidence in case we needed to take legal action." The property owners association was like a deme. "We don't have municipal ordinances out here to protect us, so we have to be organized. We are polite but persistent. We've fought off a planned fracking water disposal well near here and are currently fighting plans to put in a frack sand mine." Estimates for the sand mine's water consumption were about 2 billion gallons per year.[10]

From 2011 to 2013, 50 billion gallons of water were used for fracking in Texas. Though a big number, it's still less than 1 percent of the state's total water use. However, if you change the scale of analysis, fracking's impact is often magnified. In some rural counties on the Eagle Ford Shale, it accounts for more than 50 percent of total water use.[11] In Montague County it's about 18 percent. In Wise County it's around 43 percent.[12] One report found that 261 so-called monster wells collectively consumed twice as much water as is used by an entire rural county in Texas.[13]

North Texas has been locked in severe to extreme drought for several years and faces critical groundwater conditions. The little town of Krum in western Denton County is "running short of water" yet nonetheless agreed to sell 36 million gallons of it to Vantage Energy for fracking.[14] The story of Texas mirrors a global concern, as most of the shale plays around the world are in areas that are arid or under high levels of water stress. Fracking is increasingly a prominent competitor for scarce water resources.[15]

Then I thought back to Dr. Ireland's remark about making rain. The problem is that there is no guarantee that the new water created when methane is combusted will fall in the places where the water was extracted. It's proffering a diffuse, global response for an acute, local problem. New rain over the ocean or in the Alps won't recharge Terry's water well near St. Jo.

We made stops at three homes that day, playing on lots of farm equipment and narrowly avoiding a run-in with a huge fire ant mound. By far our favorite stop was at the home of Toby and Janelle Thompson. Toby was a tall old farmer who wore overalls and reminded me of my gentle friend Frank back on the Pennsylvania farm. Janelle was so sweet with MG as they drank lemonade and played with their dog, Blackie. They had lived in that home for many years, and I could see why; the back yard was lined with majestic crepe myrtles and sat atop a hill overlooking a lush valley. It was the most beautiful place I had seen in North Texas.

Toby and I watched Terry as he took out his sonar device. He turned it on for me to hear, and it sounded like the turn signal on a run-down pickup truck. Terry then inserted a little gray tube into a hole atop the well to take his readings. While he did that, I asked Toby how his water well had been doing. "We've been lucky," he said, "because we haven't seen much of a drop. But many of our friends 'round here have."

Then I asked the big question: "What do you think is causing the drop in water levels?"

There was a pause, and I could see that Janelle was now tuning in to our conversation. "Well . . . ," Toby said, "we have our ideas. Certainly there's a drought . . . but it's got to be the excessive pumping. You can see all the frack ponds in this area from satellite images." Then Janelle chimed in, "There used to be lots of little natural ponds around here where we would catch big bass. But they're all dried up now. I've seen entire streams go dry, leaving nothing but dead fish."

Once again, lack of monitoring perpetuates doubt about the issue. According to the director of the regional groundwater district, unlike homes and businesses that have water consumption meters, government regulators use an "honor system" for the oil and gas industry. Regulators "just don't have the manpower to check all the wells." [16]

MG AND I got Blizzards from Dairy Queen for the ride back home. As we ate, he asked, "Can they light their water on fire?"

"You mean Toby and Janelle?" I could see him nodding in the rearview mirror. "No," I said, "that's a whole other issue that is confusing for me. I mean, there's the question of how much water fracking uses, and then there's the question of whether fracking pollutes water."

"Well, does it?" MG asked.

"Does fracking pollute water, you mean?" I looked in the rearview mirror again. He nodded, licking off his spoon. "Well, I hate to say this, but the answer is kind of yes *and* no. I mean, it kind of depends on how you define the terms."

I could see him rolling his eyes . . . *Here goes Dad again. He never has a straight answer on this stuff.*

"No one denies that fracking puts lots of toxic chemicals into water," I said.

"So it does pollute water!" MG interjected.

"Sure," I said, "but we pour some toxic chemicals, like bleach for our laundry, down our drain at home, so we pollute water too."

I could see this had him thinking, so I went on. "But here is where things get murky. The wastewater from our house stays in pipes apart from other water until it is cleaned. Then it can go back into lakes and creeks. The question is whether that happens with fracking too."

"What do you mean?" he asked.

"Well," I said, trying to find the right way to put things, "the toxic water from fracking goes down a big long pipe—maybe over a couple of miles total down into the earth and then across through the shale. Then it's pumped through cracks in the pipe. Some of it stays down there. Lots of it comes rushing back up and has to go into tanks and trucks either to be reused for more fracking or to be disposed of.[17] There's really no good way to clean fracking wastewater like they clean our home's wastewater to go back into streams." [18]

"So some of it could spill?" MG said, showing remarkable patience with his dad as I struggled to think out loud.

"Yeah, that's a good way to put it! And some of it does spill. It might come out of those trucks or those big pits we sometimes see along the side of the road. We now have 17,000 gas wells around this area, but the water treatment plants that give us clean water haven't updated the chemicals they treat. Fracking is potentially introducing lots of new chemicals into the drinking water supply, but we are not looking for them." [19]

MG looked concerned about that, and I didn't want to scare him, so I said, "But I've only ever heard of one fracking spill in Denton, and that was three years ago. A gas well worker got caught dumping something into Hickory Creek. The tests only looked for a certain type of chemical, and they never really figured out what actually happened. The gas well company said they didn't cause any contamination and that whatever chemicals were found in the creek were just from people spraying weed killer and spilling oil in parking lots and stuff like that." [20]

There was a long silence. As we turned south, I saw a crow alight from the lonesome gray carcass of an oak tree and fly over a pair of lazy horses grazing in the distance. A small group of black cows was napping in the shade of a giant transmission tower. I had started to think that MG'd fallen asleep until he ventured, "So it's not a big problem?"

"Well, um . . . , " I said, unsure. I had been preparing to teach about this in the fall semester, and I had recently come to terms with this issue, which had long left me perplexed. I had a way to explain it to college students. Now I had to try with a first grader.

"Usually, I trust our country's government on questions like that. You know, Obama."

He was studying leaders in his class, and "Obama" had kind of become synonymous with our country, such as a six-year-old might conceive of it.

"But they don't really know," I went on, "because they are still doing a huge study about it. They also stopped some of their studies for no apparent reason. And, sadly, our government decided not to even regulate or monitor hydraulic fracturing. They call it the Halliburton Loophole."[21]

"What's that?"

I had to think fast. "Hmmm . . . Oh, you know when we practice tying shoes and you make the bunny ears? Those are loopholes, and the law has a loophole in it that says something like 'Normally, we watch really closely when someone pumps dangerous chemicals underground, but for the oil and gas industry we'll make an exception.' So no one really keeps track of what kind of chemicals, how much of them, and where they end up."

I could see MG curling his lip in disgust or suspicion about the loophole. He was wearing his jersey from his summer basketball league, and I thought for a second about how he'd been talking with his dad about fracking for over half his life now—from his superhero phase through the wild cat times and now on to his basketball craze.

"Then, while scientists keep studying the issue, the gas industry says that there has never been a confirmed case of hydraulic fracturing contaminating groundwater, you know, the underground lake that Toby and Janelle get their water from."

"But are they lying?"

"Well, not really, because what they mean by 'hydraulic fracturing' is something very narrow. They mean that when the toxic water is pumped into those little cracks over a mile below the surface, there is no evidence it *directly* travels through all those layers of rocks above into the aquifer." [22] I was doing my best to drive and use hand gestures to illustrate the point, wiggling my fingers upward, trying to crudely symbolize the movement of water through rocks. I then said, "Watch, we are going to drive for a mile and imagine we are driving straight up from the big cracks made in the rock during fracking to the aquifer." We drove in silence for a minute or so. After a mile, I said, "Okay, so the industry says there is no way dirty water could go through solid rock all that way."

MG said, "I think the Magic School Bus should take that trip."

"Ha! I don't think they'd do an episode where Ms. Frizzle teaches about fracking. Way too controversial. Now think about the people opposed to fracking—"

"Like Frack Free Denton?"

"Yeah, well, we define the term in a way that I think is more honest. Because fracking involves pumping that polluted water down the big long pipe. And then remember lots of it comes back up that pipe. And those pipes crack and break sometimes. That means the polluted water could leak through the pipes into groundwater. And that would be an underground spill." [23]

"Has that happened in Denton?" MG asked.

"I don't know . . . I mean, I have heard people on the outskirts of town tell me their well water went bad after fracking happened nearby. But I haven't seen any studies. It takes a long time and lots of money to research that question, and even if there seems to be a clear-cut case, debate keeps going on." [24] The best way I could think of to explain the complexity of water monitoring was to say it is decidedly *not* as simple as a pregnancy test. Given the array of chemicals and their various sources and pathways of movement, we can't really expect (as with a pregnancy test) a yes or no answer with high confidence. But I wasn't about to get into a discussion of pregnancy tests with MG.

I didn't know then that a study in the Denton area had been done and would make waves in the press two months later (just as the politics of the fracking ban heated up), showing arsenic and two chemicals used in fracking in several water wells. Of course, uncertainty remains about the source of the chemicals.[25]

"Dad," MG said, "can I play Minecraft now?"

"Yes, of course."

AFTER THEY HAVE been hydraulically fractured, gas wells cough up a good portion of the stuff that was shoved down them. An average well in Denton might spit out close to a million gallons at first (with more coming up over the life of the well). This geological afterbirth is called produced water. In addition to the fracking chemicals, produced water also contains brines, heavy metals, and naturally occurring radioactive materials (NORMs). The produced water is poured down saltwater injection wells that, hopefully, lock it away for good. Most of the produced water coming out of Denton's wells goes down the Casto injection well.

The Casto is a 13,000-foot-deep hole in the ground at the end of a bumpy dirt road surrounded by pasture in Ponder, a tiny town on Denton's western border. At the bottom of the Casto sits the Ellenburger formation, a giant geological sponge that absorbs all that waste like a beneficent god cleansing us of our sins. The Casto is just one of 680,000 waste injection wells in the country. It has disposed of 600 million gallons of toxic water in the past seven years.[26] Nationwide, over 30 trillion gallons of waste have been pumped down injection wells.[27]

Darren Groth, the head of Denton's Gas Well Inspections Division, once gave me a tour of the Casto. This was in 2012, when DAG was still helping the city revise its ordinance. Darren has a round, boyish face and a neatly trimmed goatee. Though he is friendly and usually smiling, he is also the consummate bureaucrat: always dressed in a suit and never on a first-name basis with anyone but his wife. As he drove me in his big white gas well inspection truck down the dirt road toward the Casto well, he told me all about oil and gas waste disposal. I tried to take notes on my yellow legal pad, but at this point the road got as bumpy as the Shale Voyager in the Perot Museum, and my handwriting started looking less like letters and more like a seismograph. When we got to the well, which was buzzing with activity as trucks came in to dump their contents into the large pale green tanks on-site and went back for more, I could see on the horizon, about a mile away, the neatly arranged homes of Robson Ranch and the lush green fairways of its golf course.

When the Railroad Commission permitted the Casto in 2007, it concluded the project would achieve "injection zone isolation" (the wastewater would stay where they dumped it), because there are no nearby wells "that penetrated deep enough to encounter the Ellenburger formation and thus there are no avenues for the vertical migration of fluids." However, Dan Steward, George P. Mitchell's chief engineer during the early Barnett boom, wrote that same year that "most horizontals still break out of zone to some extent and connect with Ellenburger water." [28] When the injection well was approved in 2007, there were just six gas wells within half a mile. But my map from 2012 showed fourteen gas wells that close.

The more holes drilled in the ground, the more fractures are created for water to move in unexpected ways. As one scientist put it, "There is no certainty at all in any of this. . . . You have changed the system with pressure and temperature and fracturing, so you don't know how it will behave." [29] In other words, to modify Heraclitus's famous quote about stepping in a river: you never drill into the same formation twice. In 2011, a disposal well in southern Texas suffered a "breakout" when wastewater traveled a quarter mile underground before traveling back up an abandoned oil well to leak onto a ranch. [30]

An independent investigation of injection wells discovered more than 17,000 well integrity violations nationally over just three years. [31]

And in California, fracking injection wells dumped 3 billion gallons of contaminated water into aquifers.[32] Yet despite such reports, monitoring of disposal wells is spotty, especially for the class 2 wells that handle oil and gas waste. They are often permitted in batches, meaning hundreds can be approved at once, and they are only required to be inspected once every five years.[33]

My tour with Darren was before the swarm of earthquakes in early 2014 around a similar injection well in the nearby town of Azle. There is now little doubt that injection wells can trigger induced seismic events (earthquakes caused by human activity). Induced seismicity is going to be another good test of the proactionary ethic. After the Azle earthquakes, the Railroad Commission hired a scientist, and plans were put in place to deploy more seismic monitors. The tricky part will be getting companies to disclose the frequency and intensity of their injection activities, long protected as trade secrets.[34]

Still, as one Sierra Club representative said, at least the Railroad Commission has taken the first step in admitting that a problem exists. And it took the next step by proposing concrete policy measures to fix the problem. As with any renovation, this entails taking more factors and values into account. In this case, companies seeking to operate injection wells would first need to consult with the United States Geological Survey to ascertain risks for induced seismicity.[35]

The industry's and state regulators' reactions to the fracking conundrum in Denton, however, were vastly different. Rather than sincerely acknowledge the problem and propose concrete solutions, they would offer only empty rhetoric about "responsible drilling."

A FEW DAYS AFTER my trip with MG and right before the Fourth of July parade, I got an anonymous call on my office phone. The woman on the other end of the line told me that someone was going to start circulating a petition in favor of fracking. She said the petition drive was being run by an out-of-town agency and the goal was to get 8,000 signatures to take to City Hall for the public hearing on July 15. It was just a plebiscite petition, so it would not (as with our petition) actually propose any ordinance. Rather, it was meant as a symbolic show of force.

I decided to call Peggy Heinkel-Wolfe at the *DRC* just to give her a heads-up. I thought it sounded too strange to be true, but the next day social media was buzzing about an army of petitioners that had descended on Denton. They were showing up everywhere—on the courthouse square, outside public buildings, down by the mall, and even on campus.[36]

Sharon caught one petitioner lying through the use of her spy pen that she sometimes wears clipped to her shirt pocket. He said his petition was asking companies to frack "outside the city limits."[37] Cathy had a run-in with another petitioner who told her there was no fracking currently allowed in Denton but that on July 15, City Council was going to vote to allow fracking. He told another woman that his petition was to ban fracking. One of our avid supporters called me after speaking with a petitioner and said, "They are either lying or they have no freaking clue what they are talking about." The man Cathy spoke with was from Colorado and said Denton Taxpayers for a Strong Economy was paying him $2 for every signature he collected (plus travel expenses). We later found out that the petition drive was being run by a company based in my old hometown of Colorado Springs.[38]

The Sunday before the public hearing, Breitling Energy ran a full-page ad claiming that "Denton is on the World's Stage and the Choice is Very Clear. . . . Tell city council that we support America. YES to energy independence and NO to being controlled by foreign oil." That same day, the Briggle family hopped on our bikes to grab a few groceries at the nearby Kroger. MG had been riding his own bike without training wheels for nearly two years at that point. Lulu sat in a little seat we'd ordered from Holland and mounted to the front of Amber's pink cruiser. On our way back, we were flagged down by a young man with a clipboard asking us to sign his petition.

He said he was from St. Louis and admitted he was being paid $4 per signature (the rate had apparently gone up). At first, he said that "the other petition" (the one that had been circulated by Frack Free Denton) was to ban fracking in the whole state and that his was just to ban it in the city limits. When Amber told him this was simply not true, he modified his story in an attempt to fish for the $8 he'd get if we signed his petition. Lulu just kept ringing the bell on Amber's bike. MG waited patiently in the heat, probably feeling pity for this poor guy who had somehow been foolish enough to get himself into a conversation about fracking with his mom and dad. The man obviously regretted approaching us, because other people were walking by who were likely to sign with fewer questions asked. Eventually, we signed the petition: Amber as Mickey Mouse and me as Josh Fox.

The first two speakers at the public hearing a few days later were Bobby Jones and Randy Sorrells, two of the few locals to strike it rich on fracking and cochairs of the opposing PAC, Denton Taxpayers for a Strong Economy. They approached the dais at City Hall with a plain cardboard box full of the petitions with 8,000 signatures from people (and Mickey Mouse) who "support reasonable and responsible oil and gas drilling regulations instead of a ban." They handed the box to the city secretary and ended with a rhetorical flourish, calling the proposed ban "a blindfold over the eyes of our taxpaying citizens, a rope around the neck of the local city economy, and a knife in the back for local mineral owners and American energy independence."

I had arrived at 4 p.m., about two hours before the start of the hearing, just to be sure to get a seat. By then, the main chamber at City Hall was already half full. Media trucks were filling the parking lot and setting up their big antennas in preparation for live broadcasts. City staffers were expecting a huge turnout. They hauled in extra chairs and established two overflow seating rooms in City Hall and another overflow seating area at the Civic Center next door capable of holding about three hundred people. They also established a check-in system so people could either submit written comments or be assigned a number indicating their place in the speaking order.

By the time I got there, they had already handed out sixty cards to people who wanted to have their three minutes at the microphone. I saw Alyse Ogletree, Maile's neighbor, and she smiled giddily, holding up her sign with 55 written on it in big black numbers. She couldn't wait to share her story on such a big stage.[39] Police were out in full force. Even the fire marshal was there. As Mayor Watts called the meeting to order, I turned around in my seat and counted thirteen different news crews lined up with cameras at the back of the main chamber in City Hall. The atmosphere was charged. A historic night was about to unfold.[40]

City Council had set aside the first five speaking slots for opponents of the ban, including industry leaders and elected state representatives. They gave the next five speaking slots to Frack Free Denton. It reminded me of a hockey shootout, where you try to pick five players to represent your team and score a goal. We set our lineup as follows: Maile (personal testimony), Rhonda (public health), Adam (economics), Sharon (legal dimensions), and Cathy (more testimonials and response to claims made by the industry).

After the leaders of Denton Taxpayers for a Strong Economy, the next speaker for the industry was Ed Ireland. He used his time to read a letter from Dr. Ray Perryman. The night before, the Perryman Group (funded by the industry via the Fort Worth Chamber of Commerce) had released an economic impact assessment of the proposed ban.[41] The letter Dr. Ireland read summarized the report's main claims: over the next ten years the fracking ban would cost the city $251.4 million in gross product, 2,077 person-years of employment, and $5.1 million in lost tax revenues. The school district would also lose $4.6 million, according to the report.

Tom Phillips was the next speaker. Formerly the chief justice of the Texas Supreme Court, Mr. Phillips had since taken a job as a lawyer for the Baker Botts law firm representing the Texas Oil and Gas Association. He used his three minutes to argue that the ban was either preempted by state law or an unconstitutional taking of private property. It would, he concluded, result in lawsuits with "disastrous" fiscal consequences. Mr. Phillips's testimony would later appear in every Denton resident's mailbox as a four-page glossy mailer paid for by Dr. Ireland's organization.

John Tintera, former executive director of the Railroad Commission, then spoke reassuringly, saying that state regulations for fracking were "stringent" and more than adequate. He repeated that "there has never been a documented case of groundwater contamination from hydraulic fracturing in its history."

By this point, it was clear that City Council members Kevin Roden, Dalton Gregory, and Jim Engelbrecht had come prepared to draw swords from the dais. They were less than amused that only now were state agencies and industry officials feigning concern about Denton. They didn't care when the people of Denton suffered—only when their own interests were suddenly on the line. In quick succession after Mr. Tintera's talk, Kevin got him to admit that the Railroad Commission does not have the authority to regulate setbacks and quality of life issues. Dalton got him to admit that he didn't know how often Denton's gas wells were inspected by the state. Jim asked, "Would it be all right with you if someone set a diesel generator 200 feet outside of your back door and ran it twenty-four hours a day?" The only response Mr. Tintera had to that was that there "should never be an end to conversation."

The next two speakers were our elected state officials, Senator

Craig Estes and Representative Myra Crownover. Senator Estes dismissed our concerns, saying the "technology has been used safely for fifty years." He went on to say that "sound science should drive these important policy decisions" and concluded by offering to help remedy the "compliance problem" in Denton. Representative Crownover spoke condescendingly, even suggesting fracking is about the same nuisance as a barking dog. "Mineral rights are the dominant estate," she said. "That's the law, and we have to deal with that." She also ended with a promise to help the city resolve its problems. Of course, nothing ever came of their promises.

While these five spoke, the chamber remained silent and serious. Amber was in the Civic Center, which was not bound by the same rules of conduct. They were broadcasting the public hearing on a huge screen at the front of the cavernous space. Amber texted me, saying that was the place to be—it was full and raucous, with people booing and jeering the speakers.

Maile was next. "Because of fracking," she said, "I am going to spend the rest of my life worrying about what the long-term health effects may be for my children. The nightmare never ends." If the industry wants to claim they are not causing these health problems, she said, "then *they* should have to prove it to me, not the other way around. . . . They are bullies and they have no heart." Maile concluded, "Stand on the right side of history . . . so no other mama has to worry like I do." The chamber stayed quiet, but the Civic Center burst out in the first round of applause.

After Rhonda spoke about the emerging science of fracking's health impacts, it was my turn. I told the audience to live tweet the hearing at #passtheban and then dove into my economic analysis, arguing that fracking constituted only 0.2 percent of our local workforce and that only 2 percent of the mineral wealth belonged to actual Denton families.[42] I was nervous approaching the microphone, but once I got started, my anxiety gave way to a feeling I was unaccustomed to—it wasn't just anger, it was righteous indignation. How dare they wreck our neighborhoods and then come into our town and talk down to us like we are ignorant children? In the Civic Center, the same room where three years ago a newly pregnant Amber and I sat through the first task force meeting, hundreds of people cheered, feeling emboldened now that we were on the attack.

Texas Sharon next masterfully boiled down a report that Earthworks had commissioned from Curtin & Heefner, a Pennsylvania law firm, which showed how the proposed ban was in fact legal.[43] Then Cathy played a video montage of the blowout, flaring at the purple park, and flowback fumes at UNT and the Meadows neighborhood. She read some of the more than three hundred complaints Denton residents had filed with TCEQ just that year: "piercing headache and nausea, throat irritation, and eye burning..." From the dais, Jim asked Cathy how many e-mails she had sent City Council over the years. "A thousand," she said. The mayor started asking her questions about her personal history with fracking, but Cathy broke down in tears, overcome by the stress of the night and the years of heart-wrenching fights for better regulations. As she climbed the stairs back to her seat, for the first time, the people in the main chamber broke the rules of decorum and applauded.

The hearing dragged on past midnight. At one point, seeking a bit of levity, Dalton asked one speaker (Kelli Barr, one of my graduate students), "Are you now or have you ever been an agent of the Russian government?"

At 1:30 a.m., eleven-year-old Riley Briggs finally got his turn to speak. By then, the overflow rooms were mostly empty and even the main chamber had a few empty seats. The remaining audience was exhausted, but the sight of a boy approaching the microphone reenergized the room. Riley said confidently that it was "quite past my bedtime," but he was there because we needed "to stand up against fracking."[44]

By 2 a.m. the last speaker, Riley's mom Keely had had her say. It was time for City Council to deliberate. They had two choices: either approve the ban as a city law or deny it, sending it to the November general election. Thirteen straight hours of meetings had made them exhausted almost to the point of delirium, causing the conversation to be something less than fully rational. At one point, Mayor Watts asked Anita Burgess, the city attorney, what she thought about a technical detail. "I think it is 2:30 in the morning," she replied, indicating that the time for thinking had passed.

Council member Joey Hawkins said the ban was so historic that it needed to be decided by the voters. But Kevin proposed adopting the ban right then and there, because he had a different interpretation of the ballot option in November. "This isn't going to be a citizen vote," he replied. "This is going to be national politics hits Denton with the funding of the industry. . . . That's not an even playing field, that's David vs. Goliath. If we want a fair vote, I don't think we are going to get it." Dalton agreed with Kevin, saying he was concerned that the industry-sponsored "misleading, inaccurate information" would be "magnified significantly" if the ban went to a general election.

Kevin and Dalton made such a persuasive case that for a few minutes I started to believe City Council might actually pass the ban, sparing us the agony of what was sure to be an ugly campaign. After all the council members had had their say, Chris Watts, the newly elected mayor, delivered a long, somewhat rambling speech. He aired all the ambiguities of the situation, speaking in favor of the ban in one breath and against it in the next. I felt for him and could tell he was genuinely agonizing about what to do. He repeatedly emphasized that vested rights was the core problem and said the ban wouldn't solve it. The time had arrived to make a decision, and Chris admitted he still didn't know what to do. "I guess my vote will be my vote when it happens. I won't know until I press that button." A few minutes later, he pushed the button to deny the ban as a proposed ordinance. The council voted 5 to 2 to send the ban to the November 4 ballot.

Roughly 600 people attended the hearing; 206 submitted written comments, and 104 spoke. All totaled, 73 percent favored the ban and 27 percent opposed the ban. Counting only Denton residents, those favoring the ban stood at 85 percent. Dalton would later write in a blog post, of all the ban's supporters, "90% were from Denton. None were from Russia." [45]

Part IV

RESPONSIBLE DRILLING

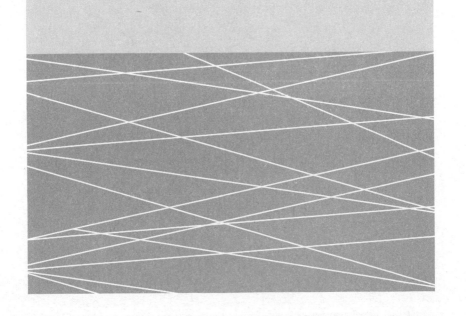

CHAPTER 9

DAVID VS. GODZILLA

I **BIKED HOME FROM THE PUBLIC HEARING SOME-** time after 3 a.m. Four hours later the Briggle clan was packed in the car on our way to Colorado for summer vacation with my parents (known as Oma and Opa, Dutch for grandma and grandpa). Things had been intense, and we needed to get away from fracking politics. But on the twelve-hour drive I couldn't stop playing the scenes of the public hearing over and over in my mind.[1]

I was energized by the hearing. And others felt the same way. A couple of my international friends and colleagues at UNT said the hearing was the night they finally decided to complete their application to become American citizens—so that they could vote for the ban on November 4.

Yet I was also troubled by the big show of force from the industry and its powerful allies. What bothered me the most was the Perryman report about the economic costs of the ban. No one had any time to read the report, let alone critically analyze it, yet its figures were parroted in nearly every story written about the hearing.[2] Kevin used his blog to fire off some quick responses. He noted that fracking was a minuscule aspect of the local economy, and royalties would continue to accrue even after the ban passed, as it wouldn't shut down existing wells. He also speculated that the ban might actually grow the economy by attracting millennials to Denton and its burgeoning high-tech sector.[3]

His analysis didn't directly confront the Perryman study, however. Supposedly on vacation, I was obsessed with debunking the outlandish claim that fracking is somehow a major aspect of Denton's economy. So, after days spent playing with Oma and Opa, I stayed up late a few nights reading the report. Lulu, who basically never sleeps, kept me company. Amazingly, even the *industry's own numbers* confirmed what Kevin had been saying. Their own report put fracking's contribution at just 0.2 percent (as in two-tenths of a percent) of the local workforce, 0.2 percent of Denton's gross product, 0.17 percent of the school district's budget, and 0.4 percent of city tax revenues. I also learned that the Perryman Group, a nonpartisan think tank, is notorious for exaggerating claims to suit their clients' preferred outcomes.[4] They often get away with this, because they use a proprietary economic formula (like the fracking industry's secret chemicals). Moreover, their own report actually showed fracking to be an economically unproductive use of land, relatively speaking. The average acre of land generates twice as much gross product as an acre of fracking, and an acre of residential development generates over four times as much tax revenue for roads and schools as an acre of fracking.[5]

I put some of these midnight findings on my blog right away. When we got back to Texas, I worked up a more polished piece, published on *Truthout* on August 11.[6]

Ten days later, I got a heart-stopping e-mail from UNT provost Warren Burggren. I had gotten to know Warren a bit over the past few years and considered him a friendly acquaintance. His note scared me stiff: "Unfortunately your recent communications regarding local government representatives has, unfairly or not, explosively developed into very serious repercussions for UNT as a whole." He asked that I meet with him and UNT president Neal Smatresk as soon as possible. I got that same sinking feeling as when the FBI came to my office. What did I do!?

My mind raced back over all my communications. I couldn't think of anything. Then it dawned on me. The tweet. It must have been the tweet. Six weeks earlier, as I was tweeting to #passtheban at the public hearing I got caught up in the drama of the night and described State Senator Estes in perhaps not the most complimentary terms, saying that he "plays the role of out of touch asshole well." It was a mistake.

That said, though, how serious could this be? It was one brief comment quickly buried in a flood of tweets that night! And surely an elected state official is called worse names than that on a daily basis.

We set up a meeting a few days later. It was boiling hot. It was the first day of class. My bike got a flat tire on the way to work. And it was my lucky thirteenth wedding anniversary. Bob Frodeman, my friend and colleague who coined the term "field philosophy," had also been asked to join the meeting. Over the past year his UNT Center for the Study of Interdisciplinarity (which made possible my work on fracking) had been defunded and eliminated. As we waited in the president's spacious lobby, we talked about the reasons for CSID's demise. Perhaps, we speculated, the heat I had drawn with local fracking politics might be partly to blame. We were trying to model ways in which humanities scholars can contribute to the university mission of community engagement. But maybe it looked less like engaged scholarship and more like troublemaking.

Always thinking strategically, Bob said, "Well, it's not often you get the ear of the president. Let's see if we can't make lemonade out of this meeting. We can maybe tell him about field philosophy." All I could do was try to breathe deeply and groan back nervously, "I just want to still have a job after this meeting."

President Smatresk and Warren popped out of another room shortly after that. They shook our hands, smiled, and steered us into a gorgeous corner office with opulent rugs and bookshelves. We sat around a big table about the size of my entire office, and without wasting time, the president told me what had happened. He had just been at a meeting with the board of regents (the governing body for the university, appointed by the governor of Texas) to discuss UNT's severe budget problems. Recently, it had come to light that UNT had, through faulty accounting practices, misappropriated roughly $80 million in state funds. The state auditor was claiming that UNT needed to repay the state, but everyone was still trying to assess the fallout and the next steps.[7]

So there was President Smatresk (who had recently been hired and inherited this financial mess) in a crucial and tense meeting with the regents. But rather than get down to business about the budget, he told us, "they only wanted to talk about one thing: a tweet." He was frustrated that precious negotiation time was wasted while he got his ear chewed off about a comment by one of his associate professors. I was floored that a tweet could derail high-level budget talks. I looked around at the faces in the room, thinking, "This is absurd. Someone say it." But all I said out loud was very simply, "I apologize."

The tension in the room lifted a bit after that. We turned the discussion to field philosophy, which the president seemed to find interesting. Still, a lot was left unspoken. Could it be that certain state officials were using the budget crisis as a way to pressure the upper administration into silencing me? The president said, "I'm sure they'd like nothing more than to see me discipline you severely. But I'm not going to do that." Though I had clearly caused him plenty of grief, he never wavered in his defense of academic freedom of speech. I had always liked working at UNT, but for the first time I was genuinely proud of my university.

As we wrapped up the meeting, Warren indicated that many influential people were consternated about how the university was leading the effort to ban fracking. Of course, UNT was doing no such thing, but it made me realize how my actions can never fully be separated from my position as a member of the university. As Warren noted, "We live in a tightly woven ecosystem." This episode also made me face up to the fact that I had made plenty of powerful enemies. Indeed, a little while later, I would receive an e-mail from a major donor (and big mineral owner in Denton) informing me that he was going to stop making donations to the university. He was going to leave UNT $1 million, but because of my actions, he took that out of his will, writing, "I cannot support an institution that would employ someone as unethical as you. You would do UNT and Denton a big favor if you were to resign."

I forwarded this note to Warren, who said it betrayed the selfishness behind a supposedly philanthropic spirit, because the donor's actions would only wind up hurting the students. Still, it made me feel awful. With field philosophy, Bob and I had been claiming that philosophers can influence actual policy processes. As the politics got nastier, we were learning about the negative impacts, what you might call the "grimpacts," of field work.

What I came to think of as "Twittergate," my own personal social media scandal, didn't slow down my writing. I did, however, mostly write anonymously from then on as a contributor to the Frack Free Denton blog. I found myself wondering: If someone is mining my tweets and moving them up the chain of command to orchestrate a staged kerfuffle designed to shut me up, what else might they be doing? I didn't want to overreact. Still, though, when I got back to my office I started eyeing the vent in my ceiling and the nooks and crannies on my bookshelves. Was there a tiny camera in here recording me? I was starting to come unhinged from the pressure of the campaign.

———

BY EARLY SEPTEMBER, the pace of the campaign was accelerating. We were trying to plan a master schedule for everything from mailers, flyers, and door hangers to billboards, newspaper ads, phone banking, and voter registration. We felt like the Wright brothers trying to get the campaign off the ground. Things kept stalling, and we couldn't quite figure out how to sustain flight and increase altitude.

On top of all that, debate season was beginning. Our first debate was scheduled for September 2 at a meeting of the Denton County Republicans. They had not yet formed an official position on the ban, so they invited our group to a forum. The other side was going to be represented by Ed Ireland and Tom Giovanetti. I learned this shortly before an impromptu interview on the Mark Davis conservative talk radio show the morning of the debate. I had not heard of Mr. Giovanetti before then, because he had not been involved in Denton's fracking politics. Listening to his segment on the Mark Davis show, I learned that Mr. Giovanetti was the president of the Institute for Policy Innovation, a free-market think tank based in Dallas. Composed and smooth-talking, he seemed to be a regular guest on the Mark Davis show.

The members of DAG decided to send me and Calvin Tillman to this debate. As a former Republican and current libertarian, Calvin, we thought, would be in a great position to speak to this audience. I spent days preparing for the debate. I could clearly talk about all the problems of fracking in Denton, but I needed to communicate how we were powerless to fix those problems. I was thinking about the third criterion for ethical real-world experiments: the technological system must be flexible so that it can be adjusted when problems are found. As Max More said, we have to be able to weed the flowers. The original innovation has to be amenable to renovation.

This third criterion is perhaps the most difficult to fulfill. There arc, of course, huge obstacles to the second criterion, monitoring; it requires money and political will. Those are still major issues when it comes to renovation. After all, those profiting from the status quo will fight renovations (which often mean regulations) perhaps even harder than monitoring. But now we face the additional challenge of facticity. By that, I simply mean the brute material existence, the givenness, of the technological system.

The Dallas–Fort Worth area, for example, was not built with sustainability in mind. Efforts to build more eco-friendly values into the metroplex are systematically thwarted by its already sprawling, inefficient, concreted, and asphalted presence in the world. I can tell you firsthand, for example, that bicyclists and pedestrians are structurally discriminated against by the built environment. If we could scrape the metroplex off the map and start from scratch knowing what we now know, it would be much easier. But we can't do that. Hegel meant something like this when he wrote that "only when the dusk starts to fall does the owl of Minerva spread its wings and fly." [8] We can only achieve understanding in hindsight, when it is too late to prescribe a new order of things. [9] It was this reading of philosophy that Bob and I were trying to change with our field work, premised as it is on the notion that philosophers can make real-time contributions. As Marx said, writing just after Hegel, "Philosophers have only interpreted the world, in various ways; the point is to change it." [10]

But change is not easy, especially in the advanced industrial capitalist society that Marx critiqued. Path dependence is one good explanation: the range of choices we can make now is limited by past decisions, even though past circumstances may no longer be relevant. Knowledge and values evolve faster and are more nimble than the material world of things. We think about technology as a symbol of the future, but as it lingers in solid form while the world around it changes, it is also always a relic of the past—even a ghost haunting us. Natural gas is often touted as a "bridge fuel" to an alternative energy future, but each additional gas well and pipeline represents further entrenchment along the trajectory defined by the existing system. More infrastructure creates more rigidity and more challenges for changing course, not to mention the fact that cheap gas delays the deployment of renewables. Fracking is a bridge to more fracking.[11]

To truly meet the first criterion of ethical innovation, informed consent, technologies must be open to continued revision in light of new knowledge. As the "informed" part of informed consent evolves (driven by science and experience), the terms of consent change.[12] One can say, "Oh, I didn't know I was agreeing to that" or "I didn't know it could be done differently."

Yet renovation, the third criterion, is not a hopelessly utopian ideal. Though facticity constrains our ability to renovate, it does not eliminate it. Think about "best available technologies." In the case of fracking, the EPA Natural Gas STAR Program, for example, encourages companies to adopt technologies that reduce air emissions, especially of methane. Examples include reduced emissions (or "green") completions, low-bleed pneumatic valves, and vapor recovery units for tanks. Water recycling is another renovation. These are all ways to account for unintended problems. In a proactionary world, policymaking is an extension not just of science but also of engineering. We don't just learn as we go along, we also tinker. The material world can be revised, albeit slowly, like changing the course of an ocean liner.

Fracking in Denton, though, was a particularly thorny challenge. DAG spent years recommending all those renovations and more to City Council as part of the ordinance revision. Some were never adopted. Yet even the ones written into the ordinance did not apply. The industry seemed completely immune from renovation. Kevin would later remark that they had the King's X, a common saying in the children's game of tag whereby one claims an untouchable status.[13]

We might define the King's X as the stubbornness of facticity alloyed with the legal doctrine of vested rights.[14] The industry was claiming that the way Denton regulated gas wells essentially guaranteed them the right to frack as much as they wanted "in perpetuity" under old rules.[15] I once sat on a panel that included Fort Worth's attorney, who winced when she described the way Denton had gotten locked into a very bad situation. So, although the implications of vested rights had not been settled by a court, there was plenty of evidence to suggest that it had one clear meaning: the industry is grandfathered under rules that were in place when they first applied for a permit or a plat.[16] That meant roughly 280 gas wells and 11,000 acres of city land were governed by rules written when Denton was much smaller and when very little was known about the hazards of fracking. That's why, as Mayor Watts once told me, "tackling vested rights is *the key* to fixing our fracking conundrum."

The vested rights law is designed to prevent the "regulatory uncertainty" that might ensue if a city could approve a project and then willy-nilly change the rules after a developer or business was already acting on the basis of that approval.[17] But this hinges on a metaphysical question about what constitutes an "ongoing project."[18] What, for example, is the ontological status of those old vertical wells in the Meadows neighborhood? When EagleRidge turned them horizontal, should that count as a new project or a continuation of the existing thing? What about refracking those wells or adding new wells to the same site; should that require a new permit under new rules? Is it like opening your garage door (a use of an existing, already permitted structure) or is it like adding a new garage (something that would trigger the need for a new permit under the latest safety codes)? And consider all those acres platted for gas well development but without any actual gas wells on them. A strict reading of vested rights would make it seem like those wells already preexist, with their facticity preloaded and fated by the design standards in place a decade or more ago.[19]

As I drove to the debate hosted by the Republican Party, my head was swimming with these philosophical quandaries. I had to convey

all that in about a minute, because I only had five minutes for my opening remarks and I also needed to touch on the health, economic, and
legal dimensions. The debate was in the all-purpose room of a community building, with the kitchen sink behind the speakers' podium
and about a hundred folding chairs arranged atop the tile floor. Calvin
and I strategized a bit in the parking lot before heading inside. On our
way in, I held the door open for one young woman wearing bright red
lipstick and cowgirl boots and carrying a big purse with the Texas flag
printed all over it. Before she walked in, though, she suddenly stopped
to read a sticker on the door. "Really? Oh, crap," she muttered and
stomped back to her car, pulling something out of her purse. I looked
at the sticker and saw it was a statement prohibiting concealed fire-
arms on the premises. I then looked over at Calvin, who just smiled,
shrugged his shoulders, and headed in. Well, I thought, at least that
would be one unarmed member of the audience.

The room quickly filled up, and we began as all such forums do
in Texas with the three *p*'s: prayers, pledges, and politicians. First,
there is an invocation from a pastor (there's always at least one church
leader in any gathering of Texans). Then there are the pledges to both
the US and Texas flags, followed by a few words of greeting from the
politicians in the audience. On that night, State Representative Myra
Crownover was in attendance. I broke out in a sweat, wondering what
role, if any, she played in Twittergate.

Ed Ireland was the first to speak. He delivered his usual even-keeled
analysis of fracking and the Perryman findings about the economic
costs of the ban. Tom Giovanetti, by contrast, launched into a rousing
paean of fossil fuels and modernity. He painted a sweeping picture
of how fracking had brought a new era of prosperity and geopolitical
power for America. He then pointed out that capitalism is premised
on private ownership of property, whereas the ban, like communism,
represented a collectivist overreach of government that would squash
personal liberties and economic growth. It was a speech that would

move any captain of industry to tears. Rhetorically, it was savvy, because it appealed to nationalism by reframing a local land use policy as unpatriotic.

In a move that I now think was unwise, I scrambled to amend my own opening remarks to match the rhetorical intensity of Mr. Giovanetti's. While it's true that I had prepared an overly wonkish presentation, it only muddled things to revise it on the fly just minutes before speaking. Nonetheless, I scribbled away as Calvin took his turn at the microphone. "Just because someone has the right to property," he said, "doesn't mean they can piss all over my property rights." He was talking about the noises, lights, and most especially the toxic chemicals he continuously found with his summa canisters in neighborhoods and playgrounds well past the fence line of frack sites. There is no right to maximize profits from one's property regardless of consequences to one's neighbors or the environment, Calvin said.[20] Hydraulic fracturing is a toxic and dangerous well stimulation technique. The mere fact that it can, under certain conditions, render mineral development economical does not suddenly confer on someone an unimpeachable right to use it.

Calvin's composure at the microphone calmed my own nerves, and I got through my five minutes in decent shape, only stumbling here and there as I tried to read some of the new remarks I had scribbled atop my prepared speech. I sat down at the front table with Ed, Tom, and Calvin, while the audience formed a long line behind a microphone to our right. Person after person pointed a finger at me (and sometimes Calvin) and hurled invectives and accusations. I had become a punching bag, and time, in its own cruel way, slowed down to a crawl. An hour of abuse seemed like an eternity. I tried my best to write down questions from audience members. I didn't get them verbatim, but I caught the gist. "You should be ashamed. You are a professor, for Christ's sake," one woman said. "You should be using facts, not scare tactics." "Well, ma'am," I replied, trying to remain calm, "some facts are scary." Another woman said, "Why did you go on Russian television twice and say you hate America?" Representative Crownover asked what I would say to a Denton resident who just started making money off of a gas well in town. "I'd say, 'Good for you, and you'll continue to make money even after we pass the ban, because it won't shut down existing wells.'" She didn't like this response and tried to interject a follow-up question, but I pointed out to the moderator that that was against the rules, so she had to take a seat.

Mr. Giovanetti heaped more fuel on the fire. He had come prepared with some quotes from my *Slate* publications and used them in an attempt to paint me both as a radical extremist and as undermining my own case for the ban.[21] He approvingly quoted my comments about proactionary innovation, saying that I couldn't possibly support a ban while advancing a proactionary approach. When I got a chance to respond, I said that proaction is an ideal and I hoped that the audience could permit a philosopher to indulge in some utopian thinking now and again. But on that night, I said, we were talking about fracking in Denton, which was far from ideal.

Afterward, Calvin congratulated me, saying he had never been to an event before where someone else was actually more hated than he was! Needless to say, I was not surprised to read a few weeks later that the Denton County Republicans had decided to officially oppose the ban. To explain why they opposed the ban, the Denton County Republican chairwoman, Dianne Edmondson, said they were in favor of "responsible drilling."[22] This was precisely what we had feared: the issue had now been tainted with partisanship. Could we win a vote in a strongly Republican city when the Republican Party was against us?

BY AUGUST, the Denton Taxpayers for a Strong Economy, working with the elite political consultants at the Eppstein Group, had fashioned a brand and a message. Their colors were black and yellow. Their mantra was "Support Responsible Drilling." And all their materials—

website, billboards, mailers, and newspaper and television ads—read "VOTE NO ON DRILLING BAN." Inside the *O* in "NO" was a stylized hand giving a thumbs-down.

Two things were most striking about this. First, they never used the word "fracking." Instead, they consistently claimed that no operator would drill a well without hydraulic fracturing, so it was a de facto drilling ban. Second, they never spelled out just what exactly "responsible drilling" means. It was left floating as an empty signifier, doubtlessly with the hope that people would project their own positive feelings onto it. Maybe it would evoke nostalgic Americana with a windswept prairie dotted here and there with the silent, nodding heads of the old oil pumpjacks like the ones Calvin's dad used to fix in Oklahoma. They never offered concrete alternatives to the currently irresponsible situation. It was just doublespeak. To me, "responsible drilling" was analogous to the way propaganda master Edward Bernays in 1929 called cigarettes "Torches of Freedom." [23] Why change your product when you can just rebrand it?

Frack Free Denton got out in front of the opponents as far as branding goes. Working with a graphic artist, we designed a brand using the patriotic red, white, and blue colors. Alan Septoff, strategic communications director at Earthworks, helped us craft a slogan: "Our Air and Water. Our Health and Safety. Our Denton." I wanted to get "our property" in there too, indicating we were making a property rights argument, but the phrase got too unwieldy. Most importantly, we branded all our campaign materials with "VOTE FOR THE BAN," and the "FOR" was always in red. The ballot language was going to ask voters if they were for or against an ordinance to prohibit hydraulic fracturing in the city limits. We knew this presented a challenge of explaining to people that voting "for" did not mean voting for fracking and that voting "against" did not mean voting against fracking. [24]

The Republicans were not the first organization to officially oppose the ban. Just a couple of days after the debate, the board of the Denton Chamber of Commerce let their membership know that they had voted to oppose the ban.[25] I got the news from Amber, who (as the owner of her own massage therapy business) had recently joined the chamber. According to Amber, none of the members were given any notice or any chance to participate in the decision by the board. It came out later that in July, three members of the US Chamber of Commerce had flown out for a meeting they'd requested with the Denton Chamber board. They were concerned about the ban's potential to set a national precedent.[26] It was clear to me they wanted the Denton Chamber board to oppose the ban, hoping to give the impression that the local business community was driving concerns that were really coming from Washington, DC.

In his statement, Chamber president Chuck Carpenter said all that was needed was "good, sound" regulation.[27] He called the ban "unilateral" and "arbitrary." I wrote in an e-mail to him that the only unilateral thing going on was the board's decision to oppose the ban without even so much as a public forum, which would be appropriate given that the City of Denton gives the chamber nearly a quarter of a million tax dollars. And the only arbitrary thing going on here was, as I explained, "the continued siting of industrial areas in neighborhoods (within 200 feet of homes) DESPITE the fact that we have a reasonable set of rules on the books already stipulating a 1,200 foot setback."

Around September 20, Denton Taxpayers for a Strong Economy sent an oversized mass mailer (about twice the size of a regular sheet of paper on glossy cardstock) to Denton residents. It so prominently featured the chamber board's stance against the ban that many people thought the mailer was paid for by the chamber itself. Kevin lambasted the board in a scathing blog post in which he very publicly announced his decision to withdraw his membership from the chamber. Calling for real solutions, not "political hackery," Kevin asked the chamber why it would waste so much energy to protect an industry that has "virtually no economic impact on our city" rather than nurturing tech start-ups or addressing Denton's growing problems of food insecurity and poverty.[28] In my own blog I said that the board had turned the chamber of commerce into an industry echo chamber.[29] They merely repeated the misleading narrative that somehow we just needed "reasonable regulations," as if we hadn't spent years trying that to no avail.

Kevin's withdrawal from the chamber was widely heralded, in particular by business owners who did not appreciate the board dragging them into such a charged political issue. I heard from an anonymous source close to the board that they were being flooded with complaints, many of them irate, about that mailer. My source told me the chamber board was trying to lie low and hope the storm would blow over.

This incident taught me two important things about the campaign. First, powerful interests were going to join the industry in opposing us. We were mobilizing a volunteer force to hand out flyers and write letters to the editor while they were buying television and newspaper ads, billboards, and mailers. The feeling of being outgunned was so overwhelming that at one point as we were counting out flyers, Texas Sharon said in a mix of exasperation and exhilaration, "This isn't David vs. Goliath. This is David vs. Godzilla!"

Second, and most importantly, the fact (and the image) of our status as the little guys might actually benefit rather than harm us. They (the industry, the chamber board, and the Republican Party) were coming across as heavy-handed bullies and outsiders. And it was backfiring on them. They were writing the David vs. Goliath narrative for us, and people began to rally around the underdog, sending us queries about how to volunteer and cheering us along on Facebook. The people of Denton are independent and smart. It wasn't hard for them to see that the other side had only the empty rhetoric of "responsible drilling," whereas our website and Facebook page provided a wealth of detailed information about the problem. The other side was talking to voters with buzzwords, while we were talking to our neighbors with real stories.

By the time of the chamber snafu in late September, we had also finally figured out how to run a campaign. Thanks to owners Ken Currin and Nicole Probst, we were able to use the GreenHouse restaurant just north of the square every Tuesday night as our command headquarters. They had their waitstaff set aside the biggest room and set up a water station for our supporters. By then, fifty or more volunteers were coming every week, ordering beers and appetizers and talking to one another around the big wooden tables. I loved coming to those meetings, because we reveled in the common cause and shared the bonds of political friendship. After a week of debates, media interviews, and block walking, it was such a good feeling to see the familiar, smiling faces in that room.

At first, we invited guest speakers and treated it like our old forums back when DAG was helping the city revise its ordinance. But thanks to Cindy Spoon, one of our strongest supporters, I came to realize that we had outlived that model. We were running a campaign now. We needed to empower and mobilize the people. They could feel this was a historic opportunity. They wanted to be a part of it. Everyone was ready to *act*. We needed a way to tap into that pent-up energy, because we weren't just educating anymore; we were going for the win.

Tara Linn Hunter was the key. At thirty-one years old, she was both a bridge to our younger supporters and experienced enough to shoulder major responsibilities. Tara Linn's dad was in the military, so she grew up partly in Texas and partly in West Germany. She also lived in Japan, Washington, Hawaii, New York, and Mexico City. Her sparkling eyes, framed by wild swirls of auburn hair, are serenely tolerant and welcoming as a result of this cosmopolitan upbringing. A self-described feminist, activist, and environmentalist, Tara Linn got her love of nature from the German mountains and her father, a conservationist who would always plant a garden as his first act whenever they moved to a new home. Though she has grown apart from her dad's faith in the Church of Christ, she still reveres him, calling him "a saint" who would never pass a homeless man without giving him something to eat. Tara Linn would later create and run an all-night "prayer vigil" in Lubbock that was actually a homeless shelter in disguise.

In New York, she studied classical singing, which led her to the music mecca of UNT, where she started training with the world-renowned opera singer Jennifer Lane. Not surprisingly, she was soon also volunteering on the weekends to feed the poor and hungry. But although Tara Linn had lived in full health all around the world, suddenly in Denton she developed severe asthma and became dependent on nebulizers and inhalers. She joined the Occupy movement and started working on air quality issues. In 2009, she joined Cathy to protest the wells near the purple park. Over the next few years, she worked off and on with DAG. In February of 2014, she spearheaded the social media launch of Frack Free Denton. Across the summer, DAG and Tara Linn drifted apart for no good reason . . . just the usual pressures and oddities that make any grassroots movement so fragile. But by September, as the engine of our campaign kept stalling, it was clear we needed her to play a leading role. When she took over as the volunteer coordinator for Frack Free Denton, it felt like all of a sudden we didn't just have supporters; we had an army.

Tara Linn transformed our GreenHouse meetings on Tuesday nights from lectures into command central. She was a quick study, learning from the political experiences of Rhonda as well as Sharon and others at Earthworks. She also benefited from Zac Trahan and his team at Texas Campaign for the Environment, who consulted with us about block walking (or canvassing) strategies. I was usually a bit late, arriving at the meetings on my bike, cursing the heat of the Texas summer as ninety-degree days stretched all the way through September. By the time I would get there, dripping in sweat, the room would be buzzing like a real campaign headquarters. Some people would be congregating at one table to talk about writing letters to the editor. At another table, others would be discussing phone banking or fund-raising.

The biggest crowd was always huddled around Tara Linn, who would be surrounded by boxes of campaign materials and sign-up sheets. She had painstakingly divided the entire city into little zones and printed out maps with assignments for our squadron of volunteers. Some of them would go door to door and talk to residents about the ban. The following week, they would report back to Tara Linn with any e-mail addresses or phone numbers they had collected, as well as requests for yard signs. She would then stay up late entering that information into our growing database of supporters.

At first, we had a flyer, and Carol invented a way to roll them, rubber-band them, and attach them to doorknobs. Despite being over sixty years old, Carol would flit through entire neighborhoods in the mornings (while it was still cool), leaving flyers in her wake. Later, we had door hangers made with different information and a little cut-out to make it easier to put them on doors. Donations kept coming in through check and via PayPal on our website, but we still could only afford two modest mailers, whereas the opposition would eventually send out over a dozen mailers, each more than twice the size of ours. But what we lacked in funds we made up with heart. By the time the campaign was over, our volunteer corps had put at least one piece of literature on nearly every door in Denton, reaching over 30,000 households spread across sixty-two square miles!

I spent most days working with the other members of the DAG Board on higher-level strategy. So it was a nice change to walk into these meetings and transition from general to ground trooper. I enjoyed chatting with the volunteers while I waited to get my assignment from Tara Linn. Each week, she'd hand me a stack of flyers and a photocopy map of a neighborhood in Denton with two or three streets marked with a pink or yellow highlighter. My task each weekend would be to get flyers on the doors of the homes marked on the map.

The week we started this process, I volunteered to actually knock on doors and not just hang literature. I joined fellow volunteers Oren Bruton, Susan Vaughn, and Michael Hennen at the purple park, and we each took one of the long north-south streets in that neighborhood—Thomas, Hillcrest, Stanley, and Ector. It was ridiculously hot for late September, and I found the whole experience physically and emotionally exhausting. The first door I knocked on got slammed in my face by a woman who said, "My family all works for oil and gas!" She said "oil" the way many Texans do, sounding more like "ole." Another man just pointed to a huge "No Trespassing" sign near his front door, scowled at me, and said, "Ya know what that means?" I actually wasn't sure, because I thought he would put a "No Soliciting" sign up if he didn't want to talk to people. But he didn't seem to be in the mood to debate the finer points of signage, so I just scurried out of there before he exercised his Second Amendment rights.

To be fair, most people were receptive, even supportive. One old woman in a flowing white sundress was out weeding her front lawn, slow and deliberate. I spoke to her as she hunched over. Glancing over at me without even standing up, she just said, "Well, if that's what we need to make things safe, then I'll vote for the ban." Another woman confessed to me she wanted to vote for the ban but felt hypocritical, because she had mineral rights elsewhere in Texas. I told her about my family's oil wells in Louisiana and how I would feel if people around there were asking for a ban. In the end, she said I had convinced her and had her vote. As rewarding as these conversations were, those three hot hours were not an experience I wanted to repeat. Afterward, when I met Susan, Michael, and Oren in the gazebo at the purple park to debrief, they were all energized from knocking on doors. I was just grateful other people were willing to do that job!

From then on, I devoted my Saturday mornings to handing out flyers—no more knocking on doors. MG would occasionally join me, and he'd take one side of the street while I took the other. We'd make a race out of it to see who could finish first. This was around the time several media outlets ran a major story about the budding fracking boom in China.[30] It gave me mixed feelings as I watched MG running with flyers between big, comfortable American homes. I thought about how fortunate we were—who could begrudge the Chinese for aspiring to our standard of living? And who could blame them for hoping fracking for natural gas would offset their deadly coal pollution problem? Then again, I wondered about kids like MG in China who might wake up to find fracking in their city or village. When problems arose, would they be able to hand out flyers with their dads too?

I HAD A DEBATE rematch with Tom Giovanetti on October 1 down at Robson Ranch, the place where the clash between drilling and development first occurred in Denton nearly fifteen years earlier. My friend Kathleen Wazny was the head of the Robson Ranch Republican Club, so she was able to arrange the forum as part of their club's official business. Knowing better what to expect this time around, I doubled my efforts to prepare for the debate. I was allowed five minutes for my opening remarks and probably spent two whole days building my PowerPoint presentation. This time I was determined to stick to my guns and not, as at the first debate, try to adjust my remarks in reaction to my opponent.

As any City Council member will tell you, Robson is the scariest place to debate in Denton, because the voters there are so well informed. Knowing this and wanting to understand the vested rights issue better, I naturally turned to Devin Taylor. He spent about an hour in my office running over the technicalities of plats, parcels, permits, and pad sites.

"Lots of plats have the magic words," he said.

"What do you mean?" I asked, having just learned from Devin that plats are basically maps that show the divisions of a piece of land and the different uses planned for it.

"They say something like 'Gas well development can occur anywhere on here.' And of course they are hundreds of acres in size."

"But some of them don't have those magic words?"

"Yes, but . . ." There was always a "but" when Devin talked to me, the same as when I spoke to MG: never a black-and-white answer. "You see, plats can be amended later and have the words removed, but that doesn't necessarily remove their legal force. And it's not clear that the industry even needs the magic words to frack anywhere on a plat. The only way out of this is to resize or evacuate the plats, but gas well operators won't give up these rights, these *vested* rights, easily."

My challenge for the debate in Robson, as with the whole campaign, was to distill complex issues into sound bites that were politically resonant—without being dishonest. The main point I needed to get across was that fracking was ongoing less than 200 feet from homes *despite* an ordinance stipulating a 1,200-foot setback distance. Devin suggested that I use my slide show to display the plats all around the city where this situation could occur again and again.

"It's so convoluted," he admitted, trying to make me feel better about my continued confusion. "I don't think anyone has a sense of what this means." But then he added, "You'd better be careful with Robson, man. Those people know their stuff. You come in there talking loose about these technicalities and they will chew you up and spit you out." Then he looked at his watch, stood up, and smiled. "I gotta run, buddy," he said as he patted me on the shoulder. "You'll do just fine!"

"Yeah, right," I said, looking at the huge file of city plats we had been discussing.

The debate was only open to Robson residents, but Kathleen said Amber could come along as my date. We had long since grown accustomed to such "romantic" outings; pretty much all of our "dates" revolved around fracking. We arrived at the elegant Wildhorse Club-house an hour early, as Kathleen requested, to check on the slides and microphone. About two hundred chairs, with gold trim and red fabric, were arranged atop the carpet in the main meeting hall. Cake, coffee, and tea were set out on a big table to the right of the podium. Amber and I milled around as the crowd trickled in, and we spoke to a reporter who was there to cover the campaign for the *New York Times*.[31]

Mr. Giovanetti breezed in just a minute before the start of the forum. Kathleen led the audience in prayer and pledges before letting the politicians in the room say a few words. This time there were close to a dozen Republican nominees there running for various elected positions, most of them unopposed. I went first, reading my script so as to use my five minutes to fullest advantage. When it was Mr. Giova-netti's turn, he just put up one slide for his opening remarks. It was the economic figures about the costs of the ban from that damned Perryman report. He didn't focus on economics in his talk, however. Instead, he strode in front of the table where I was sitting and said, "The Middle East is in turmoil right now with ISIS on the rise and all sorts of unrest. But did you notice that gasoline prices are actually going down? How could that be? It's because of fracking, ladies and gentlemen."[32]

Then he swiveled dramatically, pointing his finger at me, as he railed against the "environmentalist wackos" who were threatening an American energy renaissance. He compared the ban to a child's tem-per tantrum, with an air of condescension. Next, he singled me out for ridicule, saying I was such a radical Luddite as to suggest (quoting from one of my *Slate* articles) that Americans were "gluttonous" when it came to our lust for energy.

There was a pregnant silence as I stood there alone in front of the packed room for my two-minute rebuttal to his opening remarks. I was tempted to defend myself, but I thought about the kids on the school bus in the Meadows neighborhood and about Maile's son, Kaden, who was still getting nosebleeds. This wasn't about me, so I said, "You know, I'm going to stick to the issue here and leave the personal stuff out of it." I went on to address his misleading remarks about energy independence. "The fact is," I said, "that Denton only produces natural gas, not oil. The United States only imports a fraction of our natural gas, and basically all of that comes from Canada. The fracking ban has nothing to do with the Middle East. But it has everything to do with the health and safety of our neighborhoods."

I then talked more about the results of a summa canister air sample that Calvin had recently taken by the purple park for ShaleTest.[33] This time, he found benzene above the level considered safe for long-term exposure. "This shows," I remarked, "that fracking is not just one or two weeks. It leaves a toxic legacy at these sites for years."

Mr. Giovanetti had a copy of the ShaleTest report. He waved it in the air and then tossed it back on the table, saying that we were making up our own science and, besides, there are low levels of toxic chemicals everywhere we go.

Amber had submitted a question to Kathleen, who was moderating the Q&A, and it was the last one to be selected. "What can you tell us about the political action committees formed to lobby for and against the ban? Where does their money come from?"

Mr. Giovanetti said, "I honestly don't remember the name of the group opposing the ban." After all my hours of preparation, I was stunned that he would walk into the debate not even knowing the name of Denton Taxpayers for a Strong Economy. Heck, they had buried the town in mailers, and their logo had been all over the local paper and in sponsored ads targeted to Denton residents on Google and Facebook. Then again, Mr. Giovanetti didn't live in Denton. When it was my turn to respond, I told the audience about Frack Free Denton. In the spirit of democracy, I also supplied the name of the opposition, encouraging the people there to check out both of our websites.

Most of the retirees in the crowd probably walked into that room sympathetic to the opposition. But they also expected decency and civility and were shocked by the personal attacks against me. Kathleen said that my opponent came off as just flat rude. I wouldn't doubt it if he did more harm than good that night for his side. After all my research, I didn't get any postdebate feedback about plats, ISIS, or any other issue. I just got praise on acting "like a gentleman." It was another political lesson for me: tone, style, and rhetoric matter as much as substance.

Interpreting the David and Goliath story, Malcolm Gladwell concluded that "giants are not as strong and powerful as they seem." It turns out that Goliath likely had acromegaly. This accounts for his enormous size. But it also gave him double vision and made him profoundly nearsighted. So, although he appeared imposing and powerful to onlookers, "the very thing that was the source of his apparent strength was also the source of his greatest weakness." [34] He could not see things clearly.

I was beginning to think that was also true of the giant we were up against, namely, the industry and the groups endorsing "responsible drilling." They couldn't see how a political blitzkrieg suited for state or national politics would fail in Denton, where people respect authenticity more than power and money.

A couple of weeks after the debate and just before the beginning of early voting, a big group of Frack Free Denton supporters gathered on the downtown square (which Lulu had taken to calling simply "my party," because there was always something fun going on there). The owner of the fine arts theater on the west side of the square said we could use its marquee to hang a banner. So Tara Linn got everyone to come to a banner-raising party to celebrate this amazing chance to show our grassroots support while getting a premier (and free!) advertising space. As I was talking to Tara Linn after the banner raising, I got a text message from Kevin. Someone had leaked the results from the industry's latest poll. Unbelievably, the poll showed us in the lead.

I was about to hold my phone up for Tara Linn to see, but she wouldn't stop rattling off assignments. "We've got to get more people for the phone bank—are you coming? We've got to flyer in the Northridge area . . . and who is ordering cards for our canvassers . . . ?"

I put my phone back in my pocket, so I could quickly get out my notepad and write down yet another to-do list for the campaign. There was no time to speculate about how things might turn out. There was too much yet to be done.

CHAPTER 10

ELECTION DAY

THE NIGHT BEFORE THE ELECTION BROUGHT OUR last meeting at the GreenHouse. Tara Linn gave instructions to dozens of people who had volunteered to take shifts at the polling stations. She divided each of the twenty-six locations into three-hour slots and lined up volunteers to cover every shift from 7 a.m. to 7 p.m. at every station.

"You've got to stay behind the line marking a hundred feet from the door," she said from the front of the room while volunteers still filed in. "Be respectful, but be active and present. Remember, you are the last face they are going to see before voting. Most importantly, watch the other side like a hawk. The people the industry has hired to work at the polls will lie to voters and tell them that if they are against fracking they should vote against the ordinance. If you only say one thing, it's got to be vote *for* the ban. For! For! For! That's the key word." She was enthusiastically pumping her fist in the air with each "for!"

"Any questions?" Tara Linn smiled at the room packed full of volunteers, many of them shifting nervously in their seats. The anticipation was unbearable. Months of effort were culminating in the big day that was now upon us. I could hear people whispering speculations about how the vote would turn out. Everyone was hoping we would win, but there was also a sense of dread in the room. How devastating it would feel to invest so much time and emotion only to lose. I hadn't told anyone about the text of the industry's poll results that Kevin sent me a couple of weeks earlier. I had no way to verify those figures—and I thought they might only foster a false sense of hope or, worse, complacency.

I still didn't believe we were ahead in the polls, but I had to admit that our opponents had been acting as though they thought they were behind. In the past two weeks, they had redoubled their efforts. Our mailboxes were filled nearly every other day with mailers, and television ads seemed to play constantly on all the stations from local networks to ESPN and YouTube. We had raised enough money to run a few more ads in the *DRC*, but television was out of our budget.

Just before arriving at our last meeting, I had been going over drafts of our speeches that the DAG board members had been circulating the past few days. One e-mail, with the subject line "If we win," had our victory speech. The other e-mail was titled "If we lose." After the last ballot was cast, we were all going to gather together at our favorite bar, Dan's Silverleaf, to watch the election results, and at some point the DAG board members would get on stage to give either the "if we win" or the "if we lose" speech.

Ed and Carol brought boxes of granola bars so that our volunteers wouldn't go hungry working at the polls. Cathy brought our last remaining yard signs—enough for two at each polling location. Rhonda brought boxes full of literature to hand out. We had printed up thousands of small cards with the exact wording of the ballot language on one side (and a reminder to vote *for* the ordinance) and on the other side a picture of the fracking operation near UNT with the words "We don't even allow bakeries in neighborhoods, why this?!"

We had been using these cards throughout two weeks of early voting, and they seemed to make a big difference. It was amazing to think that despite all our work on social media, advertisements, mailers, and debates, for many people this little card might be all they really paid attention to. One picture and one phrase could be the key to this election.

Texas Sharon brought the most important item that night: umbrellas. In North Texas, rain is a rare visitor, but when it arrives, it is usually in buckets. Election Day was going to be one of those rare days. The forecast was calling for a 100 percent chance of rain lasting all day and all night.

This gave rise to more nervous speculation. About 15,000 people had cast their ballots during early voting. If this was going to be a typical election, another 15,000 or 20,000 would vote on Election Day. But rain was likely to put a damper on turnout. Whose supporters would it affect more? I remember talking to Tara Linn as we left the GreenHouse. Always the optimist, she said, "We've been helping Mother Nature, and she is returning the favor. Our supporters are the ones who are most passionate. They'll brave the rain. The other side only has money. They don't have heart."

I went home to try to get some sleep, hoping that Tara Linn was right. I hadn't slept well in over a month. Each night, I couldn't stop thinking about all the arguments I could make to help us win the vote. Thanks to a donation from local resident and supporter Phyllis Pearson Minton, we were able to afford Facebook promotions and were picking up over a thousand new Likes for our page every week.[1] We had a large audience tuning in, and I felt responsible for making the case for the ban in as many ways as possible. That meant plenty of late nights on my computer as Lulu, the night owl, played quietly with dolls or watched cartoons on my phone. In the mornings, I would wake up and quickly toggle in a panic between Facebook, my e-mail accounts, the *Denton Record-Chronicle*, and a Google search for "Denton fracking ban" (where our opponents had three sponsored ads at the top of the search results). I was looking for the famous "October surprise" from the Eppstein Group (handling the opposition's campaign). Surely they had something up their sleeve, some nuclear option that would blow us away.

My worst fear was that they would catch me doing something and gin up a scandal. For example, I had recently stepped foot on a gas well site with a journalist from Denmark and quickly ducked back in his car, thinking, "My God! There could be cameras here. I'll get tossed in jail for trespassing. It will ruin the campaign." Never before had I lived in such a heightened state of paranoia. Looking back on it, I think I was close to a nervous breakdown.

While I was lying in bed that night, Tara Linn was in a car with Gena Felker and her partner, Britt Utsler. They stayed out until 3 a.m., exhausted and delirious, putting yard signs at all of the polling locations. Tara Linn would be back up by 6:30 and on the phone serving as the brains of the operation—coordinating deliveries of cards, hot chocolate, and extra help to all of the locations.

By 6:30 in the morning, I had grabbed the sign out of my yard and was on my bike heading through the still-dark morning to my home precinct at the polling station in the North Lakes recreation center. I had signed up there for the entire twelve hours. I felt protective of my home turf. There was no way I was going to let the industry's paid poll workers lie to my own neighbors. My friend and fellow UNT colleague Pauline Raffestin was signed up for the early shift and met me there with a big rain tent that we could stand under. Unfortunately, most of the parking lot was inside of the 100-foot marker, meaning that we wouldn't really be able to talk to people. The first sprinkles started just as we got the tent assembled near Windsor, the street outside the polling location. Within minutes it was a downpour. I walked to the edge of the parking lot and tried handing a card to a voter as she scurried inside. By the time I got it in her hand it was soaked, the ink bleeding into an indecipherable blob. Then the wind turned my umbrella inside out.

I ducked back under the tent and called Tara Linn. "We can't really do anything out here," I shouted over the sound of the wind whipping in the cell phone. I could barely hear Tara Linn reply, "It's like that all over the city . . . do your best . . . just hold your yard signs and smile and give a thumbs-up to people driving by . . ." Perhaps if I had been of sound mind, I would have thought then that my presence wasn't going to make a difference. What does it matter if someone is standing there holding a sign? But that's not how I saw it. I got it lodged in my brain that I had to be there and wave at every car driving along Windsor. I couldn't see the road well enough from under the tent, though, so I stood right on the street corner. Within minutes my pants and shoes were soaked. Pauline was much better prepared, but Amber soon brought me a new pair of rain boots, a poncho, gloves, and a warm hat. I was ready for a full day of the futile but seemingly essential task of waving my yard sign at cars in the rain.

One might fairly ask at this point how this constitutes philosophy. Practically anyone could stand there, hold a sign, and give a smiling thumbs-up to motorists. For that matter, how is it philosophy to go around hanging flyers on doors? As my contribution to the ground activities had increased, especially over the past year, Amber had come to call me more and more an "activist." I never liked that title, but given what I was doing, how could I object? One might say that the only real philosophical activities were my writing and speaking when I was engaging arguments and questioning authority.

That way of thinking, though, seems to presume you can chop up the practices of your life and arrange them into discrete boxes. I don't think that is possible. I was only given a platform to write and to speak because I had been on the streets doing all sorts of "nonphilosophical" things like attending meetings at City Hall and listening to moms at Fire Station #7. Had I not put in years on the ground, for example, the *DRC* and the *Texas Tribune* would not have invited me to write op-eds during the campaign. They say you've got to pay your dues if you want to sing the blues. Well, the same is true about field philosophy. But it's not just that the grunt work allows you to have a voice. More importantly, it informs your voice. It even becomes your voice. As I look back over all my blogs, op-eds, and speaking events, I see a maturation as I became a conduit for the people and the city I had come to love. It's not that I lost my own voice. It's that the stories of others got mangled and kneaded into my ideas. What I learned in the field became the stuff of my philosophizing.

Pauline stood with me in solidarity for her three-hour shift. It's hard to explain, but we felt giddy and exuberant. The cold, wet, and wind only served to heighten a kind of mute celebration of democracy. It all felt so human, so utterly human. Pauline is French and couldn't even vote for the ban, but standing out there somehow set us vibrating on the deeper wavelengths of citizenship. No matter how vehemently people may disagree about public affairs, that very disagreement and its very vehemence betray our common core that makes us what Aristotle called *zoon politikon*, the political animal.

The industry did pay a young man and a young woman to work at the North Lakes polling location where Pauline and I were. They mostly sat in their car, though. At one point about an hour after Pauline left, the young man turned to me and said across the spitting rain, "We're going to get some lunch. We'll be back soon." They left a small pile of little cards shaped like piggy banks urging people: "Don't break the bank! Vote no on the drilling ban!" Their cards were nicer than ours and resisted the rain for a while as they sat there on the sidewalk, but eventually they got soaked and floated off into the gutter. The young man and woman never came back.

———

AMBER HAD A PIZZA delivered to me in the parking lot at around noon. By then, the grass around the tent was so saturated that my boots made little craters that quickly filled with water as I paced back and forth. In order to eat, I put my yard sign back on its little metal stand and stuck it in the ground. It slid in like a hot knife through butter. October had been so hot and dry that the soil had hardened into a rock. Cathy's husband, Ron, had laboriously muscled hundreds of signs into the ground around Denton. I had to laugh as my sign slipped so easily into the grass. I'm sure Ron wished this rain had come a month ago.

Ron and his team of volunteers had put over a thousand of our red, white, and blue "Vote FOR the Ban" signs in the yards of Denton. We easily had five times as many signs in yards around town as the opposition had. I wondered if that meant we really did have the wide-spread support to beat their money. I still wouldn't let myself believe it. But I couldn't ignore the fact that most drivers that morning had smiled and reciprocated my thumbs-up as they drove by. Standing on a street corner with my sign for twelve hours did teach me at least one useful thing: you can't predict how people will vote by the type of car they drive. Well, okay, we pretty much had a lock on the Prius crowd. But a surprisingly large proportion of pickup drivers gave me the thumbs-up. Carol had probably done the most work at the polls throughout early voting, and she had one simple rule: no profiling. You had to approach everyone, because there was no way to tell from someone's car or appearance whether he or she was inclined to vote our way.

Gena was signed up to join me around 2 p.m., which gave me two more hours of solitude standing on the corner. Outside of the occa-sional car, the only noise was the rain placking on the hood of my poncho. It was the first moment of quietude I had had in a month. It gave me a chance to think back on the recent frenzied events of the campaign.

I first thought about the swanky goat-roast fund-raiser hosted by the ethnomusicologist Steve Friedson and fiber artist Elise Ridenour in the spacious back yard of their historic home on Oak Street near UNT. They normally hosted the event to raise money for African charities, but this year they turned it into a party for Frack Free Denton. Not being a very adventurous eater, I didn't try the goat, but Lulu seemed to love it. After the mariachi band played their set, Ed got up to play drums with a makeshift band that called themselves the Frack Free Trio. Dana Isaac, the lead singer, crooned, "Well if they think they're gonna be frackin' here, they can just go frack themselves . . . ," as families milled around eating exotic foods and bidding on crafts by local artisans in a silent auction. It was so very Denton. I noticed that lots of people stood up to get closer to the stage once Ed started playing. I asked somebody what was going on, and he said, "Are you kidding me? It's not every day you get to hear Ed Soph live on the drums." It hadn't really sunk in until that moment that our treasurer—the man who had been tracking PayPal transactions and Facebook ad receipts for the past year and giving us budget reports as we sat around Rhonda's living room—was a world-class musician.

A week later, the anthropologists Lisa and Doug Henry put on another fund-raiser at their house off of South Bonnie Brae, close to several wells and the property of the two cochairs of Denton Taxpayers for a Strong Economy. They just happened to be the parents of MG's best friend, Will, and had the coolest back yard with a pool and a zip line.

Fund-raisers were crucial to our campaign needs (for example, we raised enough money at the Henrys' party to pay for another full-page color ad in the *DRC*). Maybe even more significant, though, was their social function. They solidified the bonds of a grassroots organization.

Fund-raisers, house concerts, and community bike rides were network-
ing venues where people got plugged in, empowered, and energized.

The other side had nothing like that. I'm not sure if they tried organizing and it failed or if they just never tried, figuring that they had
sufficient resources without having to bother with the groundwork
and logistics of building a movement. Denton Taxpayers for a Strong
Economy spent $1.1 million fighting the ban. Oil and gas companies
EnerVest and XTO each contributed over $200,000. Occidental
Petroleum put in $50,000. Chevron added another $150,000. Devon
Energy, the largest contributor, put in $350,000.[2] Far less than 1 per-
cent of their campaign funds came from Denton residents. And the
$1.1 million figure did not include the push polling (i.e., the phone
calls like the one Ed received asking if he'd be less likely to support the
ban if he knew Putin was behind it) and what must have been close to
another $200,000 spent by Ed Ireland's Barnett Shale Energy Educa-
tion Council. It's possible other contributions went unreported too.
As Peggy Heinkel-Wolfe at the *DRC* once told me, "If there is dark
money being spent in this town, we'll never know it." Even if there
was no other money involved, the fracking fight was thirty times more
expensive than any other campaign in the city's history.

For our part, we reported total contributions of about $76,000, mean-
ing we were being outspent roughly fifteen to one. Now, $40,000 of our
contributions came through Earthworks. We were always clear that we
asked for their support, because that's what they are good at doing—
helping communities stand up to extractive industries. Strictly speaking, Earthworks cut the checks for many campaign items including yard
signs, some newspaper ads, and the website. But Frack Free Denton was
not *funded by* Earthworks. More than 95 percent of the money Earthworks spent on Frack Free Denton originated in Denton. Dentonites
gave the money to Earthworks to give to the campaign. We couldn't say
for sure why some local residents gave to Earthworks rather than directly
to the campaign.[3] One good guess, though, is that by donating through

Earthworks, people didn't have to share their name with anybody. They could protect their identity, which is a reasonable precaution when you are fighting the most powerful industry on the planet.

Of course, the opposition pounced on this in another attempt to paint us as outside extremists. In a move that would have been hilarious if it weren't so exasperating, Denton Taxpayers for a Strong Economy issued a press release (after the first of two required campaign finance reports) asserting that 98 percent of their contributions came from individuals and businesses paying taxes in Denton.[4] Their cochair Bobby Jones went on to claim that "over 60% of the pro ban groups [sic] funding is coming from extreme liberal fringe groups out of Washington, D.C." Ed called it polishing a turd: suggesting that a group getting fat industry checks was somehow the genuinely local campaign. And I guess they were trying to promote a bit of doublethink by labeling an all-volunteer group hosting all sorts of community events as the real outsider.

At the time, I kept assuming that the opposition knew exactly what they were doing. But in retrospect, so much of what they did and said seemed badly miscalculated. Accusing us of being un-American was the best case in point. The Denton Taxpayers for a Strong Economy website, for example, asserted that I was a "far left activist" with "direct connections to a Russian Government TV entity." They said the same thing about Sharon. Sharon asked Randy Sorrells (one of the Denton Taxpayers cochairs) why they made this claim, and he had no answer. I had given one interview with a Russian television station.[5] I guess that made me a Russian. It was all so laughable that it only served to fuel our campaign; jokes about rubles, vodka, and comrades became a commonplace at our GreenHouse meetings. Rhonda felt slighted that no one accused her of being a communist, because she actually had the family ties to Russia that Sharon and I lacked.

Their mailers also seemed to backfire, often generating more skepticism than support. One mailer featured a healthy apple on one side

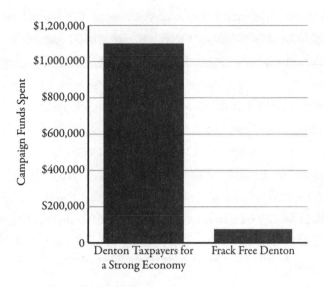

Campaign Financing for and against the Ban

When it was all said and done, the opposition outspent us fifteen to one.

labeled "responsible" and a rotten apple on the other side labeled "irresponsible." The caption read, "Denton's irresponsible drilling ban proposition will hurt our schools' financial health." Another one had a stock photo of smiling kids in a classroom and read, "Denton's drilling ban proposition will hurt our schools." Yet another one had a stock photo of a young girl on a swing on one side labeled "responsible" and an old, faded, rusty, and broken swing set on the other side labeled "irresponsible." It claimed that the ban "will hurt our parks and recreation areas."

For the most part, people were offended. How dare they claim that risking the health and safety of our kids is somehow worth the money? And Denton voters knew enough to next ask: What money?! Each time one of their mailers came out, we were able to mobilize our growing social media network to dismantle their claims. Using city data,

I showed how fracking was just 1.2 percent of our Parks and Recreation budget and how none of that revenue went to maintaining playground equipment.[6] I calculated that the ban would cost at most $20 per schoolchild out of a budget of $7,000 per child. And of course, by opening up more space for residential development, the ban might actually increase the school budget.[7] We also quickly ran an ad with a picture of one blue shoe that read, "1 shoe. The value of your child," to convey just how cheaply they wanted us to sell our kids' health and safety. People could see how the opposition's narrative about school funding just didn't add up. If fracking was so great for schools, why is Texas ranked forty-ninth in the country in funding per student?[8] And what about all those cities that don't have fracking—are their swings all falling apart?

A couple of days before early voting, their bogus petition drive came back to haunt them. Denton Taxpayers for a Strong Economy took out a full-page ad in the *DRC* listing some of the 8,000 names they had gathered back in July just before the public hearing. They were trying to show the strength of their numbers. The headline in the paper the next day, though, read, "Hot under the Collar," because several residents said their names had been printed without their permission. One retired UNT professor had signed the "poorly worded" petition in July because the paid petitioners lied to him and said it was to put fracking up for an election (rather than its true intent, which was to oppose the fracking ban). When someone called later in October asking his permission to use his name in the ad for the paper, he clearly said no twice. But that week, his name showed up in the ad. Others had a similar experience.[9]

A few days later, I ran across Kevin at Denton's big Day of the Dead festival on Industrial Street just east of the square. Fracking politics was *the* issue at the time, so we got to talking about it while Lulu ran around in her fairy outfit (actually a princess, but Amber and I are leery of the whole princess culture) and MG entertained us with the swordplay that is apparently traditional for "black dragon ninjas" such as himself.

Referring to the opposition, Kevin said, "I think these guys keep shooting themselves in the foot."

"How so?" I asked while deftly preventing Lulu from sticking a candy that had fallen on the dirt back into her mouth, distracting her with a little spider ring for her finger.

"Let me put it this way," Kevin replied. "I've got a good friend who is a hardcore libertarian. He was dead set against the ban, because he saw it as big government intruding on private property. But after all these ads and mailers . . . well, now he's voting for the ban. He's come to see your side as the people just asking for their basic rights to enjoy their homes, while the other side is actually the one with the heavy hand. Remember, there are a lot of libertarians around here, and many of them don't trust corporations any more than they trust government."

"That's a good point," I said.

"Did you get my text with the poll results?" he asked with a sly smile.

"Yeah," I said, pushing down the thought that we might actually win this thing.

AT TWO O'CLOCK, Gena arrived for her shift in the rain with me. If there were awards given for best hugs, Gena would have shelves full of trophies. In the often all too dull constellation of society, she is one of those bright stars that radiates a positive love of life. She attracts people to her like a magnet, just because you want to feel uplifted and shake off that sleepwalking everydayness of plugging along waking, eating, working, sleeping . . . or to put it more simply, as MG did after meeting Gena: "She's fun!"

It was a relief to see her, knowing that I'd have a great conversationalist to snap me out of my glassy, zoned-out trance on the street corner. But I worried about her health, because Gena has a rare form

of thyroid cancer. It's been pretty much a mystery for her and her doctors, but she has told me several times she suspects fracking may be to
blame. Of course, there is no way to know, but she had lived fairly close
to a frack site in Denton (before moving to our neighborhood), and
others (including some pets) in her old neighborhood had also become
ill. The wind had died down some, but it was still cold and she only
had a thick blue headband and a cheap red parka to keep the rain out.
She had stuck a black-and-white Frack Free Denton sticker onto her
parka. The long sleeves of her shirt were soaked within minutes. After
a while of hanging out together, I took a quick bathroom break in the
rec center. When I got back, Gena was fired up. "Oh, you should have
seen it!" she said. "This guy drove by and I gave him a thumbs-up with
my sign. He frowned and gave me a thumbs-down. So, I gave him the
peace sign. Then he flipped me the bird! And it was all in just a few
seconds like that, bam bam bam, as he drove by!" She was reenacting
the scene with great animation and sweeping arm gestures. She was
loving it—even when people flipped her off. I started to think she was
going to be just fine out there in the cold rain.

We made a kind of game out of catching every driver's eye to see
what kind of reaction they gave us. Gena thought it was funny when
people tried so hard to pretend like they didn't see us in that uncomfortable two seconds as they drove by. As we smiled and waved, we got
to talking, and she asked, "What was the deal with that empty chair
debate a while ago?" So I told her the story of one of the most stressful
moments in the campaign.

On September 23, Dianne Edmondson, the chair of the Denton
County Republican Party and an outspoken critic of the ban, invited
Frack Free Denton to take part in another debate scheduled for October 14. By then, we had debated Dr. Ireland and Mr. Giovanetti (neither of them a member of the opposing PAC or a Denton resident) on
several occasions, and we felt that since it was a local ballot initiative
we ought to have one debate that featured a local resident speaking

against the ban. So on September 24, we wrote back to Ms. Edmondson that we would participate "only on the condition that the other
side's representative is a current member of the Denton Taxpayers for
a Strong Economy's Board and a resident of the City of Denton."

We didn't hear anything until October 6, when Ms. Edmondson
said Richard Hayes, a local attorney, would represent the other side.
We asked if he was a member of the Denton Taxpayers board, and she
replied, "Yes, he is." But he was not listed on their campaign finance
filings or even on their website. When we asked for evidence of his
position, we got only silence in return.

What should we do? We called a special meeting of the DAG leadership at Rhonda's house. I felt adamantly that we had to participate
in the forum. In fact, I had never felt so passionately about anything
during the whole campaign. "If we do it, it will be two hours of our
time and then it is done. If we don't do it, it will haunt us for the rest
of the campaign. It will be the nail in our coffin. Imagine the field
day the Eppstein Group will have with a group of professors refusing
to take part in a public forum! We've been waiting for their 'October
surprise'—well, this is it! They will have the event with an empty chair
up there and make hay with it, saying we are too afraid to show up."

I looked around the room, and I could see that the other DAG board
members were hearing me out carefully. I continued, "I know they are
yanking us around here. And it obviously won't be true that we are
afraid to debate—hell, I've been to forums hosted by the Republicans,
Robson Ranch, Kiwanis Club, and the Southeast Denton Neighborhood Alliance. Cathy is doing another Republican event. Ed is doing
a forum for the League of Women Voters. But the public won't see our
refusal to participate as a principled stand against these shenanigans—
they'll just see that we aren't there."

The meeting was incredibly tense and emotional, because others felt
just as strongly in the opposite way. "No way can we do this," Cathy
said, trying to contain her rage. "We cannot capitulate to the same

dirty tricks. They are doing just what the industry always does. Rather than comply with the rules, they try changing them to suit their interests. That's the very thing we're standing up against here!"

We went round and round the issue. At one point, Cathy nearly stormed out of the house.[10] My heart was pounding; I was afraid we were about to blow everything. Eventually, it was clear we would not arrive at a consensus. I lost in a split vote, and we announced on October 10 that we would not take part in the forum, because they had failed to meet our one stipulation.[11]

The *DRC* quickly ran a letter from Ms. Edmondson, who wrote that the event was already announced and that we were "big boys and girls" who should "act like adults" and "show up or shut up." [12] She was clearly trying to start the blowback that I feared. Yet rather than argue the terms of our condition had been met, she said we had no right to "dictate" the terms. But it had obviously been a negotiation (not a matter of dictating). All along she had acted like she accepted the terms of the negotiation and even said to us via e-mail that she had met them. Apparently, that was a lie. Her complaint was *not* that we failed to uphold our end of the bargain but rather that we never had the right to bargain. But this was all after pretending for nearly three weeks that we were engaged in good-faith negotiations. It was misleading and duplicitous.

In the end, the debate went ahead as scheduled. There was indeed an empty chair up there. Denton Taxpayers for a Strong Economy and the Republican leadership were licking their chops, hoping for a media circus about how we had "chickened out" (in the words of a tweet Mr. Giovanetti sent me about this affair).[13] But then something wonderful happened. When the moderator asked if anyone in the audience would like to fill the empty chair, two of our supporters, Michael Hennen and Elma Walker, stood up and walked forward. It was especially fitting that Elma took the challenge. She is a resident of Robson Ranch and had been an original DAG member who actually left the group in 2012, saying that some of our proposals (like asking for a 1,000-foot setback distance) were too radical. And here she was in front of hundreds of people and the media, speaking for the ban!

Once again our opponents' schemes were foiled, as the only real press coverage the next day was a *DRC* article titled "Residents Step Up to Debate." [14] Elma got the last word in the article with her important remark that this was not a partisan issue. Denton is one of the most staunchly Republican counties in the nation; if their position against us carried much heft with voters, then we had no chance of winning. Elma said it well: "This isn't a Republican thing or a Democrat thing. This is a Denton thing." [15] A week went by before the start of early voting. If they were going to turn our "no show" into political capital, they would have to act fast. But nothing materialized. Their "gotcha" plan had fizzled out.

Two weeks of early voting started on Monday, October 20. Ed had used his connections in Denton's amazing music scene (where everyone is in at least two bands) to schedule a voting kickoff concert that night at the amphitheater stage in Quakertown Park. The headliner was the world-famous and all-Denton polka band Brave Combo. A local artist, Jamie Pritchett, made promotional material that read, "The Frack Free and the Brave Combo." And we spoofed the "Rock the Vote" message, by billing the concert with "Polka the Vote" and "Polka your way to the polls." They played an amazing show. Lulu danced so hard she actually fell asleep that night.

For the opening act, Tara Linn, Angie, and Nikki sang the song "Fracking Is Your Town's Best Friend" as "the Frackettes," a troupe they created to mock the industry's efforts at public relations. One of the lyrics goes: "The fumes in your air may be a bit concerning, but fracking is your town's best friend." [16] They then put on a hilarious skit with a mustache-twiddling industry executive trying to figure out how to fool the people of Denton with the aid of his giant cardboard robot "Propaganda Machine." The executive would say things like "But how do I get them to see how wonderful fracking is?" The Propaganda Machine would make whirring and clicking noises, and then a little cardboard lightbulb would appear above it with a *ding* and the robot would say in a nasally mechanical voice, "MMM. Tell them that their swing sets will fall apart without fracking." "Brilliant!" the executive would shout. The skit ended when some children from the audience ran up to the Propaganda Machine and shoved three over-sized cardboard wrenches into its mouth. Each wrench had a different word written on it: "community," "democracy," "truth." The machine conked out and fell over.

Only two polling locations were open during early voting: one at the Civic Center in the heart of the city, and the other out on the loop near the mall and the more suburban and sprawling parts of town. We had volunteers lined up to cover every minute at each location, especially after Carol, Cathy, Rhonda, and Sharon had to repeatedly report the paid industry workers to the elections officers on the first day for lying to voters. It remained hot, dry, and sunny throughout early voting, and there was almost no shade in the areas designated for poll workers. Yard signs became makeshift sunshades as we roamed the parking lots handing out our cards and monitoring what the opposition said.

Things were relatively calm at the Civic Center, but Carol had taken to calling the other location "the box," from the harsh prison confinement area in the film *Cool Hand Luke*. It was at the box that one industry worker kept shouting, "They gonna take my husband's job away!" Another man drove into the parking lot in a big burgundy van plastered with signs that said, "Vote No to Fracking Oil Ban, if you don't like American oil, get your car sold" and "You can support Middle East Oil where they slaughter men, women and children, or support the USA." [17] He walked up to Cathy, who was wearing her Frack Free Denton shirt and holding her yard sign, and asked, "Why aren't you against the deaths and mayhem in the Middle East?" Cathy backed away slowly, saying, "I'm just trying to keep our children safe."

When it was my turn to take a shift at the box, Cathy called me up and warned me to stay alert. "There are some real whackerdoodles out there." That's where I met Larry, who said he had recently lost his job and was being paid by the Denton Taxpayers for a Strong Economy to walk around the parking lot handing out those little piggy-bank cards. By then, the opposition must have learned from their polls (a tool we couldn't afford) that the lawsuit angle was their strongest one. So everything they were doing was branded with the stock photo of a piggy bank being smashed with a judge's gavel to symbolize how the ban would "waste millions of tax dollars on lawsuits." Of course, here again they were stretching the truth. Never mind that it is hardly a "waste" to defend values like health and safety; the most expensive municipal ban had cost around $200,000 at that time, and Denton's city attorney had said we would be able to tap a $4 million risk fund to defend the ban should it pass.[18]

It was at the box where I first saw the mobile billboard the industry had paid to drive up and down the streets of Denton. When others had told me about it, it sounded too weird to be true. But sure enough, there it was in the parking lot: a big truck modified to haul a billboard around that read, "Support Responsible Drilling. VOTE NO on Drilling Ban," with the thumbs-down in the "NO." I guess "responsible drilling" meant that when it is done right, gas development is safe. But surely it was not being done right in Denton, was it? If so, they needed to explain how 200 feet from homes is okay. If not, then they needed to explain what changes they would make. But they never made any sort of argument. They just kept driving a billboard around.

Sharon was with me at the time. After we were finished gaping at the mobile billboard, she asked Larry, "What, exactly, does responsible drilling mean?"

There was a long pause; then Larry said, "I suppose it means pre-
serving marine life and such."

"You mean, like, whales?" Sharon asked in disbelief.

"Yeah, I suppose," Larry said. I suddenly imagined a great T-shirt:
"Save the whales. Don't frack Denton!" I was filled with pity. I felt
sorry for Larry, who was really just doing a job. But I was also frustrated that they were sending people to the polls who had absolutely
no idea what was going on.

I watched people trickling into the polling place to vote. Some
ignored me. Others smiled or gave me a high five. One family even
asked if they could snap a picture of their kids with me. They were
celebrating not just the ban but democracy.

It made me think back to Plato's general distrust of democracy,
because there is a very cynical story you could tell. Forget the whackerdoodles. Imagine you took the average Denton voter (about sixty
years old) and gave him or her an hour, even three hours, to listen to
the most knowledgeable people on either side debate the fracking ban.
Say it was Texas Sharon vs. Ed Ireland. They could argue with each
other the whole time, each making reasonable points, and each able to
challenge the framing and the claims of the other. The voter would be
caught in a classic dilemma: dueling experts. Whom to believe? Socrates might argue the right response is aporia, that is, complete inaction
in the face of intractable doubt.

But voters, often working on the basis of less than this imagined
debate, had made up their minds and were taking action by casting
their ballot. In *The Republic*, Socrates says that the people (*demos*)
"see many justs but not real justice" and that they "believe but know
nothing of what they believe." [19] Only the philosopher, who sees the
unchanging Form of justice, truly knows. Thus only philosophers are
fit to rule. Whether they are going in there to vote for or against the
ban, they cannot possibly *know* what they are doing. Democracy is
rule by the ignorant.

One problem with this idea is that in our society the alternative to democracy is a plutocracy married to a technocracy: money and experts. They won't be ruling from some Form of justice. Rather, they will take their bearings from entrenched interests and built-in assumptions—the very things philosophers would challenge. Meanwhile, the people give in to a kind of blissful slavery, content to consume whatever goodies a relatively comfortable system doles out.

But maybe even more problematic is this whole notion that there is a Form of justice and that someone could know it with certainty. I think problems like fracking are "wicked" in the sense of being irreducibly complex and amenable to a variety of legitimate perspectives. There is no bootstrapping your way to a view from nowhere. So what do you do in the face of interminable doubt? You can freeze up and stay out of the fray. Or you can put your chips down on the side that you think has the preponderance of evidence and the best arguments going for it.

BY 5 P.M. it was getting dark. Phyllis Minton and Kate Eaton joined Gena and me for the after-work rush at the North Lakes polling site. It was still raining, and Gena and I were still going strong, hopped up on the half-dozen hot chocolates that other volunteers had delivered to us across the afternoon. Phyllis and Kate patrolled the portions of the parking lot outside the 100-foot marker, hoping to get cards into the hands of the last voters. Who knows, maybe this last wave would cast the decisive ballots. We had to keep the pedal to the metal, knowing that our friends and volunteers at other polling stations around town were doing the same thing.

I had just put cool new lights around the wheels of my bike—the front in blue and the back in red to match the colors of our yard signs. As it got darker, I set my bike on the street corner to try to catch people's attention. By 6:30, though, it was pitch-black. No one was going to see our thumbs-up in the air. I finally surrendered to the fact that my presence there was futile. So I called Amber for a ride, and Gena and I packed up the tent in the back of her car.

Amber was anxious to get to Dan's Silverleaf for the election returns party. But first we dropped MG off at a friend's house. Amber chatted with the friend's parents while I changed out of my wet clothes in their bathroom. Lulu was staying with Amber's mom, affectionately known as Mema, in a nearby hotel. Amber and I were also going to stay at a hotel that night. Things had gotten spooky enough in the last few weeks, I thought, to warrant precaution. Frack Free Denton had received a couple of borderline death threats, someone had tried to break into the front gate at Cathy's home, and in general we knew that feelings were running hot and plenty of people were angry at me. The Denton Police Department had even assigned an officer to shadow us at all of our speaking events and another officer to patrol my neighborhood. That night, in case we won and someone lost their cool, the only one home would be our cat, Shadow.

Amber and I had talked about the election, and both of us shared a deep-set feeling that we wouldn't win. It just seemed like the odds were stacked against us, given the money and endorsements on the other side. "Just remember the Alamo, babe," she told me as we walked up to our friend's house. After an exhausting day, I was certainly nervous about the results, but I felt far calmer than I'd thought I would. I felt like, as we used to say in hockey, I left it all on the ice. I did all I could have done.

Win or lose, I was proud of the community we had built and the awareness we had raised. As I slipped off my wet clothes, I thought back on some of the highlights. I thought mostly about the Frack Free Coffin Racer that Jeff McClung had built. As part of Denton's Day of the Dead Festival, there is an annual soapbox derby race down Hickory Street off the square. Dozens of organizations and businesses enter soapboxes shaped like coffins. Jeff put in weeks of his spare time crafting a beautiful wooden coffin racer adorned with our logo and "Vote

4 the Ban" stenciled on the sides. I was honored to be the pusher as we went head to head in our heat against the coffin from the Waffle Wagon food truck.

As we pulled up to the starting line, I was surprised by the size of the crowd, piled two people deep all along gates marking the straight racetrack down the hill. I wrapped up an interview with a reporter from the BBC, and camera crews were darting in front of us, giving us the kind of media buzz that naturally follows when you unleash the creativity of your supporters. We narrowly lost our heat, and I wondered whether we wouldn't have won if my back hadn't been hurting from getting such little sleep (usually on the couch). Of course, the results of that race didn't matter. What mattered was being there and really taking part in the community. Larry was there too, handing out their piggy-bank cards—but no one wants literature at the Day of the Dead festival. They want to watch the races.

I had other flashes come to mind as I put my rain boots back on. I thought about the light brigade organized by Cindy Spoon where people held up lighted letters on the square spelling "VOTE FOR THE FRACKING BAN." A young man who actually fracks wells in West Texas asked our crew why they wanted to ban fracking. When they explained there were 281 wells in town and the ban only applied in the city limits, the man said, "They're fracking in the city? That's crazy." And he held one of the letters and joined them for a while.

I also thought about the morning I worked the polling station that opened on the UNT campus for the last week of early voting (the first time ever for that site) and saw Nikki literally running after a student to get a card in her hands. And I thought about Tara Linn and Angie assembling the big drilling rig at the campus polling site every morning and breaking it down and hauling it home on the giant skateboard every evening. They had put so much chalk on the walkways around campus there was no mistaking where to vote! The opposition (once again making bogus claims, this time that the ban would increase student tuition at UNT) was clearly threatened by how well our team did at turning out the campus vote.[20]

Amber and I said good night to MG and set off to Dan's. I told Amber all about my day of thumbs-ups. One elderly man braved the rain to come out and speak to me under the tent for a while. He said that this issue had been the most contentious he'd ever seen in his entire seventy years of living in Denton. I couldn't tell how he felt about it, and I was bracing for a tongue-lashing. But he just pushed up a bit higher on his cane and simply said, "Thank you."

Probably the best story came from Cathy. A taxicab pulled up and dropped an old man off at his polling location. He had a walker and moved slowly over to Cathy. He said, "I've just had a heart attack, I've got a kidney stone I'm trying to pass, and my wisdom tooth is impacted. But I broke out of the hospital today to vote for this ban. Ain't nothing going to stop me!"

Of course, Devin, the da Vinci of Denton, who had helped me build the giant skateboard for the drilling rig, wasn't going to be at the party. He was at home on his computer, monitoring and analyzing the voting returns as they came back. Just as we passed the courthouse, only a minute from the bar where we would watch the election returns, I got a text from him that read simply, "Congrats, 58%. It is over."

I just about passed out. Were the early voting numbers out already? Was that a solid margin? I texted a furious string of questions back. By the time he replied, Amber and I were already walking into Dan's. A long, dark, and narrow bar with a big stage on the left side, Dan's is the incubator of great Denton musical talent. It is widely considered *the* venue to play. The owner, Dan Mojica, was a big supporter, so he allowed us to take the place over and it was swarming with Frack Free Denton supporters. CNN was playing on the big screen, with Wolf Blitzer breaking down the national election returns. But the place was buzzing, because everyone had pulled up their smart phones and saw the early voting results.

I saw Texas Sharon first. She was sitting at a table in front of the stage covered with packages of new socks she had bought for all the poll workers coming in from the rain with soaking wet feet. "Did you see?!" she said, holding her phone up for me to see the early voting numbers, which showed us with around 9,000 votes and the other side with around 6,000.

Suddenly feeling the toll of standing outside all day, I slumped down to the ground at her feet. Looking up at her, I asked, "What does it mean?" I felt like jumping up and down, that is, if my legs had any energy left. But maybe it was too early to celebrate. Could they come back from that deficit?

"I don't know," Sharon said, "but we can't say anything until Cathy gets here."

I agreed. Amber and I sat at a neighboring table while friends and supporters bought us drinks. We were just waiting for the official results to go up on the big screen. The bar kept filling up, now with news crews anticipating a big celebration. The whole place was like a soda bottle someone had shaken and then unscrewed the cap halfway. Calvin was out of town but keeping close tabs on the early voting returns. At around 7:40 he sent me a text that I wished I could frame: "Congrats, you're my hero."

It seemed like forever waiting for more results to come in. When I first arrived, I could see City Council members Kevin and Jim on the television screen above the bar. They were being live broadcast at a meeting a few blocks away. After they got out, they hightailed it over to Dan's to witness the events. Even George Campbell, the usually apolitical and reserved city manager, stopped by for the historic night.

Cathy was a total wreck. I finally saw her in the corner standing with Ron, who had worked the polls in his cowboy hat all day down at Robson Ranch. I told her that NPR wanted to interview her. She just looked at me, her eyes welling with tears of happiness and relief and just shook her head no, unable to get a coherent sentence out. "Okay,"

I said, "I'll field this one, but you owe me!" We hugged, and she was barely able to whisper over my shoulder, "Did we really do it?"

At that point, Dan instructed his staff to put the Election Day results on the big screen. As precincts reported and our total vote count climbed over 10,000, our margin held. I saw Ed, who had the biggest grin on his face. Knowing that people would expect a speech from us one way or the other, I said, "Should we say something?" Ed normally knows just what to do, but this time he shrugged his shoulders. "I don't know," he said. "Ask Rhonda." Rhonda was outside being interviewed amidst all the news vans and film crews. I did an interview with NPR in Cathy's place while I waited to ask Rhonda if we should make a speech. While I was being interviewed, Gena walked by. I just had to take a quick break to grab another of her famous hugs.

By then, I had lost track of Rhonda, so I headed back inside to find her. A moment later, the announcer got on the speaker.

"I've just received this written statement from Chris Watts, the mayor of Denton."

Everyone fell silent. Amber hurried over to grab my hand.

The announcer began reading Chris's statement: "As I have stated numerous times, the democratic process is alive and well in Denton. Hydraulic fracturing, as determined by our citizens, will be prohibited in the Denton city limits . . . "

The statement continued, but no one could hear it over the eruption of cheers. Overwhelmed by the victory, the noise, the lights . . . everything, I hugged Amber, and we both kept repeating in disbelief, "We did it. We did it."

I had come to see that giants are not always as strong as they appear. And shepherd boys are not always as weak as they may seem. Like David, we had a sling. It was the people. It was fitting that Election Day was rainy and windy, because for years the wind was blowing in our faces. It seemed that all the powers were aligned against us. How much easier it would have been to turn our backs and let the wind carry us away—along the comfortable avenues of apathy into the hidden corners of collective irresponsibility.

But we stood there in the wind and in the rain. Immoveable. You should look at the pictures of our volunteers on Election Day. The sun never appeared. It was night all day long. Yet their faces were glowing as if lit up from the inside. That is the light and the splendor of glory. It is the torch of a caring heart. More than anything else, I am grateful that I got to see and feel that glow.

We were the sling. We were the strength. What we saw in Denton was a victory for grassroots democracy—the kind of thing that's not supposed to happen anymore in the age of big political money. Well, it happened on that day. That's what I couldn't believe as I got on stage with Rhonda, Tara Linn, Sharon, Cathy, Ed, and Carol.

Looking back, I like to think we helped Denton to find its voice. I hope that we made it a healthier and stronger community. As frack sites proliferate across the world, I hope that those who are put in harm's way are able to find their voices too. And I hope that those with power and money will listen.

ACKNOWLEDGMENTS

THE OPPONENTS OF THE BAN knew long before the final votes were counted that they were going to lose. They were ready with their counterattack. At 9:08 in the morning after the election, while I was lying in a hotel bed nursing a hangover, attorneys from the Baker Botts law firm walked into the Denton County District Courthouse. They had come to file a lawsuit on behalf of the Texas Oil and Gas Association against the City of Denton. The industry group claimed that "a recently passed City of Denton ordinance which bans hydraulic fracturing and is soon to take effect is preempted by Texas state law and is therefore unconstitutional." That same morning, the Texas General Land Office filed a separate lawsuit against the city, arguing much the same thing.

The fracking ban also earned the ire of Governor-elect Greg Abbott, who targeted Denton in his first major speech after winning his election. "Texas is being California-ized," he said. "We're forming a patchwork quilt of bans and rules and regulations that is eroding the Texas model." He added that he has a vision where "individual liberties are not bound by city limit signs." [1] State Representative Phil King filed two pieces of proposed legislation shortly thereafter. One would require cities to get the blessing of the attorney general before putting a citizens' initiative up for a vote. The other would require cities to pay the state for any costs associated with local regulation of oil and gas development. Other bills targeting cities were also filed. Conservative politicians who normally sing the praises of small government and local control quickly sought ways to neuter municipalities. Why? Because municipalities are the only remaining seat of politics that is accountable to citizens rather than big money.

I naïvely thought things would slow down after the election, but the fight had only begun. Our focus has shifted now from passing the ban at the ballot box to defending it in the courthouse and at the statehouse. As I write this, Tara Linn and Sharon are working with Calvin to organize a bus trip down to Austin to lobby our state legislators. Cathy is working with attorneys Deborah Goldberg, Daniel Raichel, and Bruce Baizel from Earth Justice, Natural Resources Defense Council, and Earthworks, respectively. Those organizations are helping DAG as we become interveners (essentially codefendants) in the lawsuits filed against the city. Ed, Carol, and Rhonda are lining up Denton residents who are willing to sign affidavits about how hydraulic fracturing has negatively impacted them. And I am feeling guilty for taking so much time off to finish my book!

I don't know how things will turn out, but I do know that I have been educated and inspired by my journey as a field philosopher in Denton. For that, I owe a great deal to a great many people. There is something presumptuous about this book that leaves me feeling uneasy. I mean to say that this story is so much bigger than me. It touches and has been shaped by so many lives that it's not quite right for one person to try to pass it off as "his book."

Above all, I am indebted to Amber, Lulu, and MG. Amber sometimes calls me the smarty-pants in the family, but I know the real story. She possesses a powerful mind that often guided me out of muddles. She lit my path in dark times and gave me shelter from the storms. My efforts as a field philosopher would also not have been possible without the support of Barb "Mema" Volney, who tirelessly and lovingly cared for the kids while I was off at events and meeting after meeting. And I owe Oma and Opa for modeling citizenship and civility.

I am grateful for the loving struggles I shared with Cathy, Ed, Carol, Rhonda, Tara Linn, and Sharon. With his notion of field philosophy and his (now defunct) Center for the Study of Interdisciplinarity, Bob Frodeman created the conditions that made it possible for me to be involved in this story. You never pay your mentors back. You only try to pay it forward, so I suppose there's no sense in trying to catalog the millions of ways I owe Bob. I am thankful for President Neal Smatresk, Provost Warren Burggren, and other leaders on the UNT campus for defending the core value of academic freedom of speech. I am also grateful to my colleagues Keith Brown, Kelli Barr, Steve Fuller, Matt Story, Mahdi Ahmadi, Carl Mitcham, and Britt Holbrook, who helped me think through the politics of fracking and the promises and perils of field philosophy.

Kevin Roden did so much to make this story and this book possible, including reaching out to Bob and me initially to see if it might make sense for philosophers to get involved in the polis. I should add a word of thanks to Chris Watts, Jim Engelbrecht, Dalton Gregory, Pete Kamp, Mark Burroughs, Joey Hawkins, Greg Johnson, John Ryan, Anita Burgess, Darren Groth, Lindsey Baker, and everyone else with the City of Denton who worked so patiently with me and other citizens on the thorny issue of urban fracking. I also want to thank Police Chief Lee Howell and his entire team for helping to keep us all safe. And I want to thank those members of the industry who dialogued with me. I benefited from hearing perspectives that challenged my own.

Everyone at Earthworks has earned my gratitude for offering so much of their time and expertise and taking a big gamble on a ragtag group of citizens in Denton. They are truly a top-notch organization. I am especially grateful for the work of Alan Septoff and for the generosity of an anonymous local donor. I owe Charlene for our winning image. I want to also thank Calvin, Tim, and everyone else at ShaleTest. The work they do is so vital and yet so underappreciated and underfunded.

I owe a big thank you to Angie Holliday, Nikki Chochrek, and Christina Bovinette for all their smart and dedicated work. Ken Currin and Nicole Probst did so much for us by offering the GreenHouse every Tuesday night—they had the courage to stick their necks out there as supporters. I am truly humbled by the active citizens in the Meadows neighborhood, including Maile Bush, Alyse Ogletree, Kelly Higgins, Debbie Ingram, and Sandy Mattox (just down the road). The same holds true for Elma and Bruce Walker and Kathleen Wazny down at Robson Ranch.

Of course, this story would not have been possible without the efforts of volunteers who got off the couch, tapped their creative talents, and braved the heat and the rain to become active citizens contributing to the common good. I am thinking in particular of Cindy, Phyllis, Jamie, Topher, Jeff, Nancy, Ken, Misty, Mike, Michael, Susan, Pauline, Kevin, Stone, Giovanni, Fabio, Sarah, Sara, Rosemary, Laura, Batavia, John, Marc, Devin, Jennifer, Matthew, Gena, Britt, Ron, Corey, Heidi, George, Brooke, Shelly, Steve, Elise, Lisa, Doug, Hatice, Dan, Merrie, Keely, Riley, Lyndi, Kate, Marshall, Phyllis, Harrison, Vicki, Selina, Virginia, Adam, Tyler, RayAnne, Kathleen, Andrea, Christina, Emily, Karen, Kate, Katie, Todd, Anyah, Benjamin, Maureen, David, Pam, Val, Alex, Matt, Oren, and many more. I also want to give a big thank you to the artists and musicians of Denton for rallying behind the cause.

I am especially indebted to the Think Write Publish program funded by the National Science Foundation and based at the Consortium for Science, Policy, and Outcomes at Arizona State University. Lee Gutkind and David Guston, the leaders of this program, provided me with the tools, experiences, and connections I needed to write creative nonfiction. There is no way this book happens without their guidance. I know parts of this book won't pass Lee's infamous "yellow test," but I hope it is nonetheless something of a testament to the success and importance of his program.

Finally, I want to thank everyone at Liveright for so ably and warmly welcoming me to the world of trade publishing. I am particularly grateful to Justin Cahill for his judicious comments and generous spirit as he shepherded me through this project. The book has benefited so much from his insight and guidance. I am also so thankful for India Cooper's sharp and meticulous edits on the manuscript. It has been an honor and a joy to work with such a dedicated editorial team. I only wish every author could be so lucky.

NOTES

Introduction

1. I borrow this term from Jacques Ellul's 1963 *The Technological Society*. The technological wager is a concept that has since been used elsewhere, including in the 2012 Nuffield Council report *Emerging Biotechnologies: Technology, Choice and the Public Good*.

2. As one engineer told me while I was researching this book, "Engineering is somewhere between science and art. We are precise and careful, yes, but we also take the leap. You could never get a physicist to build a laser, because they'll want to stop and ask why cesium is behaving this way or that way, whereas at some point we [engineers] don't care about that, we're just trying to get nature and machines to do work for us." I owe much of my thinking here to my mentor and friend Carl Mitcham.

3. There are interesting points of overlap between my theory and the ideas in William McDonough and Michael Braungart, *The Upcycle: Beyond Sustainability—Designing for Abundance* (New York: North Point Press, 2013).

4. Thomas Kaplan, "Citing Health Risks, Cuomo Bans Fracking in New York State," *New York Times*, December 17, 2014, http://www.nytimes.com/2014/12/18/nyregion/cuomo-to-ban-fracking-in-new-york-state-citing-health-risks.html?_r=0.

5. For more on the idea of engineering as social experimentation, see Mike

Martin and Roland Schinzinger, *Ethics in Engineering*, 2nd ed. (New York: McGraw-Hill, 1989).

6. I have in mind something like Karl Popper's idea of "piecemeal social engineering" (from *The Poverty of Historicism*, 1957) as small-scale, incremental changes made in light of ever evolving knowledge and experience (rather than large-scale changes based on some grand narrative or presumption that we have it all figured out in advance of taking action in the world).

7. For a striking example, see the fracking boom in North Dakota. Deborah Sontag and Robert Gebeloff, "The Downside of the Boom," *New York Times*, November 22, 2014, http://www.nytimes.com/interactive/2014/11/23/us/north-dakota-oil-boom-downside.html?_r=0.

8. "The Proactionary Principle," http://www.extropy.org/proactionaryprinciple.htm. In 1957 in Communist China, Chairman Mao Zedong coined the phrase "let a hundred flowers blossom" (the origin of the now more common "let a thousand flowers bloom"). He invited the Chinese intelligentsia to criticize the government, claiming that progress happens when "a hundred schools of thought contend." Of course, Mao used it as a trap to lure out dissidents, many of whom were executed after voicing their criticisms. You'll see that I think More's proactionary principle is also mostly just a smoke screen for a system (in this case, corporate capitalism) that is actually opposed to the ideal.

9. There is a rich philosophical history behind this idea of proaction that I only touch on a bit in later chapters. For more on this, see Steve Fuller and Veronika Lipinska, *The Proactionary Imperative: A Foundation for Transhumanism* (London: Palgrave Macmillan, 2014).

10. For gas wells, see Energy Information Administration, http://www.eia.gov/dnav/ng/hist/na1170_nus_8a.htm; for oil production, see Robert D. Blackwill and Meghan L. O'Sullivan, "America's Energy Edge: The Geopolitical Consequences of the Shale Revolution," *Foreign Affairs*, March/April 2014, http://www.foreignaffairs.com/articles/140750/robert-d-blackwill-and-meghan-l-osullivan/americas-energy-edge.

11. M. Bamberger and R. E. Oswald, "Impacts of Gas Drilling on Human and Animal Health," *New Solutions: A Journal of Environmental and Occupational Health Policy* 22, no. 1 (2012): 51–77.

12. Robert B. Jackson et al., "The Environmental Costs and Benefits of Fracking," *Annual Review of Environment and Resources* 39 (October 2014): 327–62, Review in Advance posted August 11, 2014, doi:10.1146/annurev-environ-031113-144051.

13. See Carl T. Montgomery and Michael B. Smith, "Hydraulic Fracturing: History of an Enduring Technology," *Journal of Petroleum Technology* 62, no. 12 (2010): 26–32.

14. Cyrus Sanati, "Behind Schlumberger's Smith Deal: A Big Gas Bet," *New York Times*, February 22, 2010, http://dealbook.nytimes.com/2010/02/22/behind-schlumbergers-smith-deal-a-big-gas-bet/?_r=0.

15. Jerry Zremski, "As Environmental Debate Rages over Fracking, People in Western Pennsylvania Express Dread," *Buffalo News*, May 17, 2014, http://www.buffalonews.com/home/as-environmental-debate-rages-over-fracking-people-in-western-pennsylvania-express-dread-20140517.

16. A former Mobil Oil executive shares this characterization. See Brian Nearing, "Former Mobil Oil Exec Urges Brakes on Gas Fracking," *Albany Times Union*, April 22, 2014, http://www.timesunion.com/business/article/Former-Mobil-Oil-exec-urges-brakes-on-gas-fracking-5422292.php.

17. In Brian O'Keefe, "Exxon's Big Bet on Shale Gas," *Fortune*, April 16, 2012, http://tech.fortune.cnn.com/2012/04/16/exxon-shale-gas-fracking/.

18. John Dewey, "The Need for a Recovery of Philosophy," in *Creative Intelligence: Essays in the Pragmatic Attitude*, ed. John Dewey (New York: Holt, 1917), 3–69.

19. I am indebted to Bob Frodeman for helping with much of this introductory section on field philosophy.

20. Robert Frodeman, *Geo-Logic: Breaking Ground Between Philosophy and the Earth Sciences* (Albany: SUNY Press, 2003).

21. See Kaitlin Toner Raimi and Mark R. Leary, "Belief Superiority in the Environmental Domain: Attitude Extremity and Reactions to Fracking," *Journal of Environmental Psychology* 40 (December 2014): 76–85.

Chapter 1. THE CITY AND THE SHALE

1. John Tamny, "Don't Fear 'Fracking,' Fear the Horrid Illusion That Revived Fracking," *Forbes*, July 6, 2014, http://www.forbes.com/sites/johntamny/2014/07/06/dont-fear-fracking-fear-the-horrid-illusion-that-revived-fracking/. And here are the data for the decline in domestic oil production, from the US Energy Information Administration: http://www.eia.gov/dnav/pet/hist/LeafHandler.ashx?n=PET&s=MCRFPUS2&f=A.

2. You can find a video of his speech here: https://www.youtube.com/watch?v=-tPePpMxJaA.

3. I later published this as a book defending President George W. Bush's bio-ethics council: Adam Briggle, *A Rich Bioethics: Public Policy, Biotechnology, and the Kass Council* (Notre Dame, IN: University of Notre Dame Press, 2010).

4. You can find the animation here: http://www.eia.gov/todayinenergy/detail.cfm?id=2170.

5. John Browning et al., "Barnett Study Determines Full-Field Reserves, Production Forecast," *Oil and Gas Journal*, September 2, 2013, http://www.beg.utexas.edu/info/docs/OGJ_SFSGAS_pt2.pdf.

6. See Russell Gold and Tom McGinty, "Energy Boom Puts Wells in America's Backyards," *Wall Street Journal*, October 25, 2013, http://www.wsj.com/news/articles/SB10001424052702303672404579149432365326304?tesla=y.

7. Stories like this of (real or perceived?) municipal powerlessness in the face of the oil and gas industry are very common. For another example from Nordheim, Texas, see Julie Dermansky, "Nordheim: A Texas Town Facing a Toxic Future," DeSmogBlog, January 30, 2014, http://www.desmogblog.com/2014/01/30/nordheim-texas-town-facing-toxic-future?fb_action_ids=10153791281770533&fb_action_types=og.likes&fb_source=aggregation&fb_aggregation_id=288381481237582.

8. Peggy Heinkel-Wolfe, "Reflection of Watts," *Denton Record-Chronicle*, May 19, 2013, http://www.dentonrc.com/local-news/local-news-headlines/20130519-reflection-of-watts.ece.

9. See Lowell Brown, "Robson Seniors Take On Drilling," *Denton Record-Chronicle*, June 1, 2011, http://www.dentonrc.com/local-news/south-denton-headlines/20110601-robson-seniors-take-on-drilling.ece.

10. For a write-up of some of these early events, see Lowell Brown, "Drilling Raises Some Unease," *Denton Record-Chronicle*, August 27, 2011, http://www.dentonrc.com/local-news/special-projects/gas-well-drilling-headlines/20110827-drilling-raises-some-unease.ece.

11. For a natural history of the Barnett, see Robert G. Loucks and Stephen C. Ruppel, "Mississippian Barnett Shale: Lithofacies and Depositional Setting of a Deep-Water Shale-Gas Succession in the Fort Worth Basin, Texas," *AAPG Bulletin* 91, no. 4 (April 2007): 579–601.

12. "Natural Gas Was City Fuel First in 1912," *Denton Record-Chronicle*, February 3, 1957.

13. US Department of Energy (DOE), *DOE's Unconventional Gas Research Programs 1976–1995: An Archive of Important Results*, January 2007, http://www.netl.doe.gov/kmd/cds/disk7/disk2/Final%20Report.pdf.

A good overview of the origins of unconventional gas in the United States can be found in Zhongmin Wang and Alan Krupnick, "A Retrospective Review of Shale Gas Development in the United States," Resources for the Future, http://www.rff.org/RFF/documents/RFF-DP-13-12.pdf.

14. Norman Carlisle and Jon Carlisle, "Project Gasbuggy," *Popular Mechanics*, September 1967, 104, 105, 222.

15. Carl T. Montgomery and Michael B. Smith, "Hydraulic Fracturing: History of an Enduring Technology," *Journal of Petroleum Technology* 62, no. 12 (2010): 26–32.

16. A. B. Yost and W. K. Overbey Jr., "Production and Stimulation Analysis of Multiple Hydraulic Fracturing of a 2,000-ft. Horizontal Well," Society of Petroleum Engineers Gas Technology Symposium, June 7–9, 1989.

17. Obituary, *New York Times,* July 26, 2013, http://www.nytimes.com /2013/07/27/business/george-mitchell-a-pioneer-in-hydraulic-fractur ing-dies-at-94.html?pagewanted=all&_r=0.

18. For more on this story and others important to the development of frack-ing, see Gregory Zuckerman, *The Frackers: The Outrageous Inside Story of the New Billionaire Wildcatters* (New York: Portfolio, 2013).

19. Dan Steward, *The Barnett Shale Play: Phoenix of the Fort Worth Basin, a History* (Fort Worth: Fort Worth Geological Society Press, 2007).

20. This quote is transcribed to the best of my ability from a talk given by Dan Steward, available on YouTube here: https://www.youtube.com/watch?v =NuTKZjUYNOw.

21. Geologists and engineers knew about and had tapped into the Barnett prior to the drilling of C. W. Slay #1 in 1981, but these were exploratory opera-tions. The term "discovery well" means the first well to commercially pro-duce oil or gas from a formation.

22. Michael Shellenberger, "Interview with Dan Steward, Former Mitchell Energy Vice President," for the Breakthrough Institute, December 12, 2011, http://thebreakthrough.org/archive/interview_with_dan_steward_for.

23. See Steward, *The Barnett Shale Play*, and Gregory Zuckerman, "Breakthrough: The Accidental Discovery That Revolutionized American Energy," *Atlantic*, November 6, 2013, http://www.theatlantic.com/business/archive/2013/11/ breakthrough-the-accidental-discovery-that-revolutionized-american-energy/281193/?single_page=true. Russell Gold in *The Boom: How Fracking Ignited the American Energy Revolution and Changed the World* (New York: Simon and Schuster, 2014) seems to correctly attribute this momentous discov-ery to the S. H. Griffin #4 well. By contrast, Zuckerman attributes it to the #3 well. Texas Railroad Commission data for these two wells show the #3 producing

massive quantities of gas in February 1998, but then quickly trailing off. The data for #4 show big numbers for that well starting later, in August of 1998, but the high level of production is sustained for a much longer period of time.

24. Anthony Andrews et al., "Unconventional Gas Shales: Development, Technology, and Policy Issues," Congressional Research Service, October 30, 2009, http://www.fas.org/sgp/crs/misc/R40894.pdf.

25. There are many articles out there about Mitchell, but a good overview of his life can be found at http://www.georgepmitchell.com. For the comparison, see, David Blackmon, "George P. Mitchell: A Visionary Life," *Forbes*, July 30, 2013, http://www.forbes.com/sites/davidblackmon/2013/07/30/george-p-mitchell-a-visionary-life/.

26. Jennifer Hiller, "Texas Oil Production Moving Up in World Ranking," *Houston Chronicle*, April 23, 2014, http://www.houstonchronicle.com/business/energy/article/Texas-oil-production-moving-up-in-world-ranking-5422511.php.

27. Jeff McMahon, "Fracking Insiders See No End to Boom," *Forbes*, May 4, 2014, http://www.forbes.com/sites/jeffmcmahon/2014/05/04/fracking-insiders-see-no-end-to-boom/.

28. Nelson D. Schwartz, "Boom in Energy Spurs Industry in the Rust Belt," *New York Times*, September 8, 2014, http://www.nytimes.com/2014/09/09/business/an-energy-boom-lifts-the-heartland.html?_r=0.

29. Jaeah Lee and James West, "The Great Fracking Forward: Why the World Needs China to Frack Even More," *Wired*, September 18, 2014, http://www.wired.com/2014/09/great-fracking-forward/.

30. From Prof. Colter Ellis's presentation at the 2014 conference "The Implications of Hydraulic Fracturing for Creating Sustainable Communities," Binghamton University, State University of New York, April 10–11, http://www.binghamton.edu/tae/sustainable-communities/cid-conference/.

31. See Rivka Galchen, "Weather Underground: The Arrival of Man-made Earthquakes," *The New Yorker*, April 13, 2015, http://www.newyorker.com/magazine/2015/04/13/weather-underground.

Chapter 2. CLOSING THE BARN DOORS

1. Adam Briggle, "Fracking? Not in My Back Yard (or Yours)," *The Conversation*, April 23, 2013, http://theconversation.com/fracking-not-in-my-back-yard-or-yours-13185.

2. I am grateful to Kevin Roden for clarifying this story for me.

3. See subchapters 5, 6, and 22 of the Denton Development Code, available here: https://library.municode.com/index.aspx?clientId=14239&stateId=43&stateName=Texas.

4. John A. Barton, deputy executive director of the Texas Department of Transportation, "Presentation on TxDOT's Energy Sector Task Force," October 23, 2012, available at http://ftp.dot.state.tx.us/pub/txdot-info/energy/102312_txdot_presentation.pdf.

5. Cathy McMullen's YouTube account has several videos of this, including this one of drilling near the purple park: https://www.youtube.com/watch?v=OXzKI6WWAuA.

6. Here is a video I shot during fracking near UNT's football stadium: https://www.youtube.com/watch?v=msL7nfabTC4.

7. For FLIR camera footage of this process near UNT's football stadium, see ShaleTest's video: https://www.youtube.com/watch?v=WJDMP5dgP5c&list=UUu_QDf79tZsmPIFONegzaRA.

8. For an example of a leak, see this ShaleTest video taken at a gas well in the western part of Denton: https://www.youtube.com/watch?v=X5FnkFvBwfU&index=10&list=UUu_QDf79tZsmPIFONegzaRA.

9. This example is taken from reports made available by Chesapeake Energy at http://www.chk.com/investors/pages/reports.aspx.

10. See Adam Briggle, "Fracking," in *The Encyclopedia of Science, Technology, and Ethics*, 2nd ed., ed. J. Britt Holbrook (New York: Gale, 2014).

11. The database (for PSE Healthy Energy Network) is here: http://www.psehealthyenergy.org/site/view/1180. Concerned Health Professionals of New York, "Compendium of Scientific, Medical, and Media Findings Demonstrating Risks and Harms of Fracking," July 10, 2014, http://concernedhealthny.org/wp-content/uploads/2014/07/CHPNY-Fracking-Compendium.pdf.

12. In a landmark case, a jury awarded the Parrs nearly $3 million for damages and nuisance caused by Aruba. This was not a direct recognition that fracking caused their health problems, but it was about as close as one can get given how difficult it is to prove a health claim (we'll return to this in later chapters). See Mose Buchele, "Texas Family's Nuisance Complaint Seen as Win Against Fracking," National Public Radio, May 2, 2014, http://www.npr.org/2014/05/02/308796539/texas-familys-nuisance-complaint-seen-as-win-against-fracking.

13. Suzanna Andrews, "A Texas Rebel's Fight for Her Land," *More*, Septem-

ber 2013, http://www.more.com/news/personalities/texas-rebels-fight-her-land.

14. Peter Gorman, "Evolution of a Rebel," *Fort Worth Weekly*, May 15, 2013, http://www.fwweekly.com/2013/05/15/evolution-of-a-rebel/.

15. Sharon's blog is at http://www.texassharon.com.

16. Lowell Brown, "Tests at Well Show Toxins," *Dallas Morning News*, March 20, 2010, http://www.dallasnews.com/incoming/20100320-Tests-at-well-show-toxins-7844.ece.

17. Ian Urbina, "Insiders Sound Alarm amid a Natural Gas Rush," *New York Times*, June 25, 2011, http://www.nytimes.com/2011/06/26/us/26gas.html?pagewanted=all&_r=0.

18. I first sketched this idea in Adam Briggle, "It's Time to Frack the Innovation System," *Slate*, April 11, 2012, http://www.slate.com/articles/technology/future_tense/2012/04/george_p_mitchell_fracking_and_scientific_innovation_.html.

19. The report was posted here: https://docs.google.com/file/d/0B4DaWHJom04lUEFxWWROQW5nWFE/edit?usp=sharing.

20. Lowell Brown, "Caution Is Their Watchword," *Denton Record-Chronicle*, November 25, 2011, http://www.dentonrc.com/local-news/local-news-headlines/20111125-caution-is-their-watchword.ece.

21. This quote is from an e-mail exchange with Dr. Ireland.

22. John Siegmund, "Natural Gas a Blessing," *Denton Record-Chronicle*, May 15, 2012, http://www.dentonrc.com/opinion/columns-headlines/20120515-john-siegmund-guest-column.ece.

23. Adam Briggle, "Residents Must Weigh In," *Denton Record-Chronicle*, March 24, 2012, http://www.dentonrc.com/opinion/columns-headlines/20120324-adam-briggle-residents-must-weigh-in.ece.

24. Adam Briggle, "Uneasy Alliance Forms," *Denton Record-Chronicle*, January 14, 2012, http://www.dentonrc.com/opinion/columns-headlines/20120114-adam-briggle-uneasy-alliance-forms.ece.

25. Lowell Brown, "Mayoral Candidates Debate Drilling Issues," *Denton Record-Chronicle*, April 28, 2012, http://www.dentonrc.com/local-news/local-news-headlines/20120428-mayoral-candidates-debate-drilling-issues.ece. There was also a third candidate, Donna Woodfork, who only collected 8 percent of the vote.

26. Subchapter 22 of the Denton Development code, 35.22.4.F.2, at https://library.municode.com/index.aspx?clientId=14239&stateId=43&stateName=Texas.

27. E. E. Schattschneider, *The Semisovereign People: A Realist's View of Democracy in America* (New York: Holt, Rinehart and Winston, 1960).

28. Peggy Heinkel-Wolfe, "City Approves Drilling Ordinance," *Denton Record-Chronicle*, January 16, 2013, http://www.dentonrc.com/local-news/local-news-headlines/20130116-city-approves-drilling-ordinance.ece.

29. Joe Fisher, "History, Sprawl, Legacy Wells Put Fracking Focus on Texas Town," *NGI's Shale Daily*, April 10, 2014, http://www.naturalgasintel.com/articles/98018-history-sprawl-legacy-wells-put-fracking-focus-on-texas-town.

30. You can find her report (with Jessica Gullion and Naomi Meier, December 2, 2011), "A Perspective on Health and Natural Gas Operations: A Report for Denton City Council," here: http://www.scribd.com/doc/76392377/A-Perspective-on-Health-and-Natural-Gas-Operations-A-Report-for-Denton-City-Council.

Chapter 3. SHALE VOYAGES

1. Chapter 92 of Texas's Natural Resources Code states: "It is the intent of the legislature that the mineral resources of this state be fully and effectively exploited and that all land in this state be maintained and utilized to its fullest and most efficient use."

2. Around this time, we changed the group name from Denton Stakeholder Drilling Advisory Group to Denton Drilling Awareness Group, but we kept DAG for the acronym. After the ordinance passed, our mission changed from one of advising policy to cultivating public awareness.

3. Adam Briggle, "The Religiosity of the Fracking Debate," Science Progress, September 6, 2012, http://scienceprogress.org/2012/09/the-religiosity-of-the-fracking-debate/.

4. Dalton Gregory, "Here Is My Stand on the November 2014 Ballot Propositions," November 1, 2014, http://www.daltongregory.com/here-is-my-stand-on-the-november-2014-ballot-propositions/.

5. I don't mean to suggest things are as simple as a dualist "pro" and "anti" perspective. There is diversity within the warring camps. For a good taxonomy of perspectives on environmental issues, see John Dryzek, *The Politics of the Earth: Environmental Discourses*, 2nd ed. (Oxford: Oxford University Press, 2005).

6. Lynn Bartels, "'Tea Party of the Left' Wages Ferocious Battle over Fracking,"

Denver Post, July 19, 2014, http://www.denverpost.com/news/ci_26180743/ tea-party-left-wages-ferocious-battle-over-fracking. We can also see the party lines muddled to some degree over the fracking ban vote in Denton on November 4, 2014: Peggy Heinkel-Wolfe, "Was the Frack Ban Vote Red or Blue?" *Denton Record-Chronicle* blog, November 13, 2014, http:// newsgatheringblog.dentonrc.com/2014/11/was-the-frack-ban-vote-red-or-blue.html/.

7. Dan Boyce, "To Resolve Feud over Fracking, Colo. Democrats Turn to Plan C," National Public Radio, August 11, 2014, http://www.npr.org/ 2014/08/11/339611053/to-resolve-feud-over-fracking-colo-democrats -turn-to-plan-c.

8. Steve Fuller, "Ninety-Degree Revolution," *Aeon Magazine*, October 24, 2013, http://aeon.co/magazine/living-together/right-and-left-are-fading-the-future-is-black-and-green/.

9. Paul Ehrlich, *The Population Bomb* (San Francisco: Sierra Club Books, 1968).

10. Simon's views were given a more recent formulation in Bjørn Lomborg, *The Skeptical Environmentalist: Measuring the Real State of the World* (Cambridge: Cambridge University Press, 2002).

11. Ed Regis, "The Doomslayer," *Wired*, February 1997, http://www.wired.com/ wired/archive/5.02/ffsimon.html.

12. John Tamny argues that the outcome of this bet was mostly dictated by changes in the strength of the dollar, but he doesn't dispute the soundness of Simon's theory of scarcity. John Tamny, "Don't Fear 'Fracking,' Fear the Horrid Illusion That Revived Fracking," *Forbes*, July 6, 2014, http://www .forbes.com/sites/johntamny/2014/07/06/dont-fear-fracking-fear-the-horrid-illusion-that-revived-fracking/.

13. John Browning et al., "Barnett Study Determines Full-Field Reserves, Production Forecasts," *Oil and Gas Journal*, September 2, 2013, http://www .beg.utexas.edu/info/docs/OGJ_SFSGAS_pt2.pdf.

14. Daniel Yergin, *The Quest: Energy, Security, and the Remaking of the Modern World* (New York: Penguin Books, 2012), 717 quoted. For a good precautionary critique of Yergin's "techno-dynamism," see Michael Klare, "Have the Obits for Peak Oil Come Too Soon?" TomDispatch blog, January 9, 2014, http://www.tomdispatch.com/blog/175791/.

15. Alan Petzet, "World Hydrocarbon Supply 'Relatively Boundless,' SEG Told," *Oil and Gas Journal*, September 23, 2013, http://www.ogj.com/ articles/2013/09/world-hydrocarbon-supply-relatively-boundless-seg-told .html.

16. Bill McKibben, "Global Warming's Terrifying New Math," *Rolling Stone*,

July 19, 2012, http://www.rollingstone.com/politics/news/global-warmings-terrifying-new-math-20120719?page=2.

17. The industry has reached upward of seven miles, but the average length in Denton is still less than one mile.

18. Graeme Wood, "Re-Engineering the Earth," *Atlantic*, July/August 2009, http://www.theatlantic.com/magazine/archive/2009/07/re-engineering-the-earth/307552/.

19. Brantley Hargrove, "Fear and Fracking in Southlake," *Dallas Observer*, November 24, 2011, http://www.dallasobserver.com/2011-11-24/news/fear-and-fracking-in-southlake/.

20. John Tierney, "Betting on the Planet," *New York Times Magazine*, December 2, 1990, http://www.nytimes.com/1990/12/02/magazine/betting-on-the-planet.html.

21. Ed Davies and Fitri Wulandari, "Indonesian Tin Industry Hits a Slump," *New York Times*, October 21, 2008, http://www.nytimes.com/2008/10/21/business/worldbusiness/21iht-tin.1.17131102.html?_r=0.

22. I'm grateful to Darren Groth for explaining this to me.

23. Information from the Texas Railroad Commission's GIS map, http://wwwgisp.rrc.state.tx.us/GISViewer2/.

24. This is an estimate worked up for me by engineers and economists on campus, comparing gas well production data with figures from the UNT wind turbines.

25. These figures come from my estimates based on the data posted about these wells on fracfocus.org.

26. See Shelley Goldberg, "Fracking Opens Up an Entire New Industry," *Wall Street Daily*, September 15, 2014, http://www.wallstreetdaily.com/2014/09/15/fracking-sand/.

27. Alison Sider and Kristin Jones, "In Fracking, Sand Is the New Gold," *Wall Street Journal*, December 2, 2013, http://online.wsj.com/news/articles/SB10001424052702304868404579194250973656942.

28. The video is available at https://www.youtube.com/watch?v=WJDMP5dgP5c.

29. These data are available from the Denton Central Appraisal District, https://www.dentoncad.com/.

30. Randy Lee Loftis, "Fracking at Front and Center in Perot Museum's Energy Hall," *Dallas Morning News*, November 25, 2012, http://www.dallasnews.com/entertainment/visitors-guide/perot-museum/headlines/20121125-fracking-at-front-and-center-in-perot-museums-energy-hall.ece.

31. Rachel Carson, *Silent Spring* (1962; Boston: Houghton Mifflin, 2002), 20.

32. John A. Barton, deputy executive director of the Texas Department of Transportation, "Presentation on TxDOT's Energy Sector Task Force," October 23, 2012, http://ftp.dot.state.tx.us/pub/txdotinfo/energy/102312_txdot_presentation.pdf. Here you can see that it takes nearly 1,200 trucks to bring a single well into production, so I multiplied this by the 17,000 wells on the Barnett Shale.

Chapter 4. FRANKENSTEIN VS. FAUST

1. S. Turner, "The Conservative Disposition and the Precautionary Principle," in *The Meanings of Michael Oakeshott's Conservatism*, ed. C. Abel (Exeter, UK: Imprint Academic), 204–17.
2. M. Oakeshott, "On Being Conservative," in *"Rationalism in Politics" and Other Essays* (London: Methuen, 1962), 168–92, 169 quoted. Here is another telling line: "Innovation entails certain loss and possible gain, therefore, the onus of proof, to show that the proposed change may be expected to be on the whole beneficial, rests with the would-be innovator" (172).
3. Good information on the precautionary principle can be found on the website of the Science and Environmental Health Network, http://www.sehn.org/precaution.html. The European Environmental Agency has also produced two influential reports titled *Late Lessons from Early Warnings: Science, Precaution, Innovation*. The 2013 report can be accessed here: http://www.eea.europa.eu/publications/late-lessons-2. The most prominent formulation was adopted as Principle 15 of the Rio Declaration from the 1992 United Nations Conference on Environment and Development: "In order to protect the environment, the precautionary approach shall be widely applied by States according to their capabilities. Where there are threats of serious or irreversible damage, lack of full scientific certainty shall not be used as a reason for postponing cost-effective measures to prevent environmental degradation."
4. I think this is the main message of some of the leading engineers and designers now. For example, see William McDonough and Michael Braungart, *The Upcycle: Beyond Sustainability—Designing for Abundance* (New York: North Point Press, 2013).
5. Adam Briggle, "Should Cities Ban Fracking?" *Slate*, December 24, 2012, http://www.slate.com/articles/technology/future_tense/2012/12/longmont_co_has_banned_fracking_is_that_a_good_idea.html.

6. The city implemented a moratorium on gas well drilling permits in 2014, which tellingly also admits that "significant and compelling environmental and land use and compatibility concerns" remain even after the multiyear ordinance revision process. The moratorium can be found at: http://www .cityofdenton.com/home/showdocument?id=19817.

7. New York State Department of Health, "A Public Health Review of High Volume Hydraulic Fracturing for Shale Gas Development," December 2014, http://www.documentcloud.org/documents/1382539-new-york-depart ment-of-health-report-on-fracking.html#document/p3.

8. Tina Casey, "Fracking Ban-memtum Builds on Both Sides of the Pond," CleanTechnica, March 3, 2014, http://cleantechnica.com/2014/03/03/ momentum-for-fracking-bans-builds-in-britain-and-usa/. The Los Angeles case was a drafted ordinance that was not voted into law.

9. Karl Marx, *Manifesto of the Communist Party* (1848), chapter 1.

10. A systematic study of over 1,700 policy issues concluded that America is basically an oligarchy: "Economic elites and organized groups representing business interests have substantial independent impacts on U.S. government policy, while average citizens and mass-based interest groups have little or no independent influence." Martin Gilens and Benjamin Page, "Testing Theories of American Politics: Elites, Interest Groups, and Average Citizens," *Perspectives on Politics* 12, no. 3 (Fall 2014): 564–81.

11. Sandra Steingraber, "Safe Hydrofracking Is the New Jumbo Shrimp," *Huffington Post*, June 4, 2012, http://www.huffingtonpost.com/sandra-stein graber/safe-hydrofracking_b_1520574.html.

12. I believe that John's thoughts are sympatico with those outlined by Thorstein Veblen in his 1921 treatise *The Engineers and the Price System*.

13. Herbert Marcuse, *One-Dimensional Man* (Boston: Beacon Press, 1964).

14. Lynn Bartels, "'Flat Earth' Fracking Ad on 'Radical Activists' Insults Some Voters," *Denver Post*, July 25, 2014, http://www.denverpost.com/election 2014/ci_26212650/flat-earth-fracking-ad-radical-activists-insults-some.

15. Cass Sunstein, "The Paralyzing Principle," *Regulation*, Winter 2002–3, 32–37, http://object.cato.org/sites/cato.org/files/serials/files/regulation/2002 /12/v25n4-9.pdf.

16. See the film *Pandora's Promise*, http://pandoraspromise.com/.

17. This line is uttered by Raskolnikov in *Crime and Punishment*.

18. And it would be good, as Ivan Illich does, to pair this proactionary Prometheus with his precautionary brother Epimetheus, meaning "hindsight."

19. I am indebted to Steve Fuller and Keith Brown for helping me think through

the archetype and lineage of the proactionary spirit and its contrast with Frankenstein.

20. Goethe, *Faust*, act 1, scene 3, lines 1225–37, from the A.S. Kline 2003 translation, http://www.poetryintranslation.com/PITBR/German/FaustIScenes ItoIII.htm.

21. See F. A. Hayek, *Individualism and Economic Order* (Chicago: University of Chicago Press, 1948).

22. Cass Sunstein, "Throwing Precaution to the Wind," *Boston Globe*, July 13, 2008, http://www.boston.com/bostonglobe/ideas/articles/2008/07/13/throwing_precaution_to_the_wind/?page=full.

23. Jacques Ellul, *The Technological Order* (1963), quoted in *Philosophy and Technology*, ed. Carl Mitcham and Robert Mackey (New York: Free Press, 1983). Ellul writes, "We are launched into a world of an astonishing degree of complexity; at every step we let loose new problems and raise new difficulties. We succeed progressively in solving these difficulties, but only in such a way that when one has been resolved we are confronted by another. Such is the progress of technology in our society."

24. William James, "The Will to Believe," section 7, http://www.gutenberg.org/files/26659/26659-h/26659-h.htm.

25. Carl Cranor, *Legally Poisoned: How the Law Puts Us at Risk from Toxicants* (Cambridge: Harvard University Press, 2011).

26. Rainer Lohmann, Heather M. Stapleton, and Ronald A. Hites, "Science Should Guide TSCA Reform," *Environmental Science and Technology* 47, no. 16 (2013): 8995–96, http://www.ecori.org/storage/Lohmann%20et%20al%20TSCA%20Viewpoint%20EST'13.pdf.

27. J. Britt Holbrook and Adam Briggle, "Knowledge Kills Action: Why Principles Should Play a Limited Role in Policymaking," *Journal of Responsible Innovation* 1, no. 1 (2014): 51–66.

28. Adam Smith, *Wealth of Nations* (1776), book 4, chapter 2.

29. Ibid.

30. You can find this same argument in modern form in Vannevar Bush, *Science—the Endless Frontier* (Washington: USGPO, 1945), http://www.nsf.gov/od/lpa/nsf50/vbush1945.htm, and in Michael Polanyi, "The Republic of Science," *Minerva* 1 (1962): 54–74, https://www.missouriwestern.edu/orgs/polanyi/mp-repsc.htm.

31. Adam Smith, *Theory of Moral Sentiments* (1759), chapter 1.

32. For a more rounded account of the philosophical lineage of proactionary thought, we'd need to include Joseph Priestley and his vision of science lead-

ing to technology leading to heaven on Earth. See Steve Fuller, *Humanity 2.0: What It Means to Be Human Past, Present and Future* (New York: Palgrave Macmillan, 2011).

33. John Locke, *Second Treatise of Civil Government* (1690), chapter 5, sections 34, 42.

34. Thomas Frank, "Free Markets Killed Capitalism," *Salon*, June 29, 2014, http://www.salon.com/2014/06/29/free_markets_killed_capitalism_ayn_rand_ronald_reagan_wal_mart_amazon_and_the_1_percents_sick_triumph_over_us_all/.

35. Philip Mirowski, *Science-Mart: Privatizing American Science* (Cambridge: Harvard University Press, 2011).

36. Naomi Oreskes and Erik Conway, *Merchants of Doubt: How a Handful of Scientists Obscured the Truth on Issues from Tobacco Smoke to Global Warming* (New York: Bloomsbury, 2011).

37. Goethe, *Faust*, act 2, scene 5.

38. Many of these ordinances were drafted by the Community Environmental Legal Defense Fund. See http://www.celdf.org/section.php?id=401. We spoke with CELDF as we were debating the ban. I appreciated their input and do respect their approach, but ultimately we decided it wasn't right for us.

39. Broadview Heights, Ohio, Community Bill of Rights, http://www.celdf.org/downloads/Broadview_Hts_OH_CBOR_2012.pdf.

40. Adam Briggle, "Tilting at Gas Wells," *Truthout*, November 27, 2013, http://truth-out.org/opinion/item/20183-tilting-at-gas-wells-whats-the-best-way-to-defend-your-community-from-fracking.

41. By treating it as a business practice, this language had the benefit of offering a way out of the vested rights loophole, which only applied to land use (not business practice) policies.

42. Ivan Illich, *The Rivers North of the Future* (Toronto: House of Anansi Press, 2005).

Chapter 5. PLAYING NICE IN THE SANDBOX

1. Jim Fuquay, "Azle Crowd Frustrated by Lack of Answers About Quakes," *Fort Worth Star-Telegram*, January 3, 2014, http://www.star-telegram.com/2014/01/02/5457081/azle-meeting-on-earthquakes-yields.html.

2. See her collected articles at Shale Stories, http://peggyheinkelwolfe.com/shale-ouevre/.

304

NOTES TO PAGES 121–128

3. Peggy Heinkel-Wolfe, "Group Seeks Ban on Fracking," *Denton Record-Chronicle*, February 18, 2014, http://www.dentonrc.com/local-news/local-news-headlines/20140218-group-seeks-ban-on-fracking.ece.

4. For "living in hell," see Edward Brown, "Compressing Agony," *Fort Worth Weekly*, May 14, 2014, http://www.fwweekly.com/2014/05/14/compressing-agony/2/. For more descriptions like this, see Suzanne Goldenberg, "Fracking Hell: What It's Really Like to Live Next to a Shale Gas Well," *Guardian*, December 13, 2013, http://www.theguardian.com/environment/2013/dec/14/fracking-hell-live-next-shale-gas-well-texas-us. "We felt like prisoners in our own home" are the words of Kelly Higgins at a City Council meeting. See the short video I made in 2014 titled "Denton: A Fracking Hell," https://www.youtube.com/watch?v=Lx5ZrZIba6A.

5. TXsharon, "Fracking Executive Confirms: Homeland Security Thinks Fracktivists are Terrorists," *Daily Kos*, November 15, 2013, http://www.dailykos.com/story/2013/11/15/1255839/-Fracking-executive-confirms-Homeland-Security-thinks-fracktivists-are-terrorists#.

6. This episode got a mention in Juliet Eilperin, "As Eco-terrorism Wanes, Governments Still Target Activist Groups Seen as Threat," *Washington Post*, March 10, 2012, http://www.washingtonpost.com/national/health-science/as-eco-terrorism-wanes-governments-still-target-activist-groups-seen-as-threat/2012/02/28/gIQAA4Ay3R_story.html. See also Adam Federman, "We're Being Watched," *Earth Island Journal*, Summer 2012, http://www.earthisland.org/journal/index.php/eij/article/we_are_being_watched/.

7. He said, "We have several former psy-ops folks that work for us at Range because they're very comfortable in dealing with localized issues and local governments. Really all they do is spend most of their time helping folks develop local ordinances and things like that." You can find audio clips here: http://www.texassharon.com/psyops/.

8. See Allan Brandt, "Racism and Research: The Case of the Tuskegee Syphilis Study," *Hastings Center Report* 8, no. 6 (1978): 21–29.

9. The Nuremberg Code is available here: http://www.hhs.gov/ohrp/archive/nurcode.html. The Belmont Report is available here: http://www.hhs.gov/ohrp/humansubjects/guidance/belmont.html.

10. Gregory Zuckerman, *The Frackers: The Outrageous Inside Story of the New Billionaire Wildcatters* (New York: Penguin, 2013).

11. The idea of engineering as a real-world experiment requiring informed consent from stakeholders can be found in Mike Martin and Roland Schinzinger, *Ethics in Engineering*, 4th ed. (New York: McGraw-Hill, 2005).

12. The SUP can be found here: http://www.cityofdenton.com/home/show document?id=17310.

13. For more on fracking and developers, see Michelle Conlin and Brian Grow, "U.S. Builders Hoard Mineral Rights Under New Homes," Reuters, October 9, 2013, http://www.reuters.com/article/2013/10/09/us-usa-fracking-rights-specialreport-idUSBRE9980AZ20131009.

14. Concerned Health Professionals of NY, "Compendium of Scientific, Medical, and Media Findings Demonstrating Risks and Harms of Fracking (Unconventional Gas and Oil Extraction)," July 10, 2014, http://concerned healthny.org/wp-content/uploads/2014/07/CHPNY-Fracking-Com pendium.pdf, 70 quoted. The database that accompanies this report can be found here: http://www.psehealthyenergy.org/site/view/1180.

15. Zain Shauk and Bradley Olson, "Colorado Drillers Show Sensitive Side to Woo Fracking Foes," Bloomberg, August 27, 2014, http://www.bloomberg .com/news/2014-08-28/fracking-foes-force-some-oil-drillers-to-tread-lightly.html.

16. See Timothy Riley, "Wrangling with Urban Wildcatters: Defending Texas Municipal Oil and Gas Development Ordinances Against Regulatory Takings Challenges," *Vermont Law Review* 32 (2007): 351–407, http://law review.vermontlaw.edu/files/2012/02/riley.pdf.

17. Ibid.

18. Matthew Fry, Adam Briggle, and Jordan Kincaid, "Fracking in a Texas Community: Who Benefits?" *Ecological Economics*, forthcoming.

19. TXsharon, "Denton Resident Who Owns Minerals Supports Fracking Ban," Bluedaze blog, July 14, 2014, http://www.texassharon.com/2014/07/14/denton-resident-who-owns-minerals-supports-fracking-ban/.

20. Marie Baca, "Forced Pooling: When Landowners Can't Say No to Drilling," ProPublica, May 18, 2011, http://www.propublica.org/article/forced-pool ing-when-landowners-cant-say-no-to-drilling.

21. See Tom Wilber, *Under the Surface: Fracking, Fortunes, and the Fate of the Marcellus* (Ithaca: Cornell University Press, 2012).

22. Brian Grow, Joshua Schneyer, and Anna Driver, "The Casualties of Chesapeake's 'Land Grab' across America," Reuters, October 2, 2012, http://www.reuters.com/article/2012/10/02/us-chesapeake-landgrab-substory-idUSBRE8910E920121002.

23. Baca, "Forced Pooling."

24. Ian Urbina and Jo Craven McGinty, "Learning Too Late of the Perils in Gas Well Leases," *New York Times*, December 1, 2011, http://www.nytimes

.com/2011/12/02/us/drilling-down-fighting-over-oil-and-gas-well-leases. html?ref=us. Chesapeake Energy would later slash the royalties through a scheme that involved creating a gathering pipeline company that charged more for the gas and then deducting those costs from the royalty payments. For an in-depth analysis of this contractual legerdemain, see Abrahm Lustgarten, "Chesapeake Energy's $5 Billion Shuffle," ProPublica, March 13, 2014, http://www.propublica.org/article/chesapeake-energys-5-billion-shuffle?utm_campaign=sprout&utm_medium=social&utm_source =twitter&utm_content=1394706160.

25. For more on this story, see Molly Redden, "The Amish Are Getting Fracked," *New Republic*, June 6, 2013, http://www.newrepublic.com/article/113354/ energy-companies-take-advantage-amish-prohibition-lawsuits.

26. The map can be found here: http://www.rrc.state.tx.us/data/online/gis/ index.php.

27. See Martin Heidegger, "The Question Concerning Technology" (1954), in *"The Question Concerning Technology" and Other Essays*, trans. William Lovitt (New York: Garland, 1977).

28. The industry even elects Texas Railroad Commission members for their "man of the year" awards. See http://www.timesrecordnews.com/news/2013/ may/07/commissioner-is-oils-man-of-the-year/.

29. Controversies about local control extend beyond oil and gas issues. Indeed, some of the first battles between municipalities on one hand and industry and state government on the other hand had to do with local smoking ordinances. When cities started restricting and banning smoking in public places, the tobacco industry worked through state governments in an effort (now largely a failure) to preempt or trump local control. One of the groups leading the charge against preemption across many issues is Grassroots Change (http://grassrootschange.net/). I thank Mark Pertschuk for this information.

30. Huber Smutz, "Urban Oil Drilling—The City's Viewpoint," paper presented at an American Petroleum Institute Conference, May 1965.

31. See Timothy Riley, "Wrangling with Urban Wildcatters: Defending Texas Municipal Oil and Gas Development Ordinances Against Regulatory Takings Challenges," *Vermont Law Review* 32 (2012): 351–407, http://lawreview .vermontlaw.edu/files/2012/02/riley.pdf.

32. Joseph De Avila, Mike Vilensky, and Russell Gold, "New York Communities Can Ban Fracking: Court Rules," *Wall Street Journal*, June 30, 2014, http://online.wsj.com/articles/new-york-towns-and-cities-can-ban-fracking-court-rules-1404145435.

33. Marie Cusick, "Pennsylvania Supreme Court Strikes Down Controversial Portions of Act 13," NPR Pennsylvania State Impact Report, December 19, 2013, https://stateimpact.npr.org/pennsylvania/2013/12/19/state-supreme-court-strikes-down-act-13-local-zoning-restrictions/.

34. Scott Rochat, "Boulder County Judge Strikes Down Longmont Fracking Ban," *Denver Post*, July 24, 2014, http://www.denverpost.com/news/ci_26209898/boulder-county-judge-strikes-down-longmont-fracking-ban.

35. This was for an episode of *America Tonight* on Al Jazeera America: https://ajam.app.boxcn.net/s/r4g1bpq2c6xq7aj869do.

36. Thomas Hobbes, *Leviathan*, part 2, chapter 29, paragraph 21.

37. I get this history from Gerald Frug, "The City as a Legal Concept," *Harvard Law Review* 93, no. 6 (April 1980): 1057–154.

38. This includes energy companies and offices of attorneys general. See Eric Lipton, "Energy Firms in Secretive Alliance with Attorneys General," *New York Times*, December 6, 2014, http://www.nytimes.com/2014/12/07/us/politics/energy-firms-in-secretive-alliance-with-attorneys-general.html.

Chapter 6. IN THE BALANCE

1. They went strangely silent later in the fall after UNT reprimanded them for illegally using an official UNT logo in one of their memes. It featured the UNT football stadium and trademark eagle and said, "Like if you agree fall means football in north Texas."

2. Albert Borgmann, *Technology and the Character of Contemporary Life* (Chicago: University of Chicago Press, 1984).

3. This isn't unique to fracking; just think of a chemical factory or high-voltage electricity transmission towers. This also shows how "local" communities are not that local after all. They are made possible by global commodities and transnational corporations. That's another tough question with local control of shale gas: Can communities really act as energy sovereigns when they are dependent on the grid?

4. Jean-Jacques Rousseau, *Discourse on the Origin and Basis of Inequality Among Men* (1755). You can find a good online version here: http://www.constitution.org/jjr/ineq_04.htm. The quote is from part 2. As the iconoclast Catholic philosopher Ivan Illich wrote, "Needs are far more cruel than tyrants." Ivan Illich, *The Rivers North of the Future* (Toronto: House of Anansi Press, 2005), 103.

5. Langdon Winner, "Artifact-ideas and Political Culture," *Whole Earth Review* 73 (Winter 1991): 18–24.

6. Carolyn Thomas, "Why Big Pharma Now Outsources Its Clinical Trials Overseas," Ethical Nag, July 10, 2011, http://ethicalnag.org/2011/07/10/clinical-trials-outsourced-oversea/.

7. This sentiment often gets tangled up with patriotism. Several folks railed against the proposed ban because they thought it would hurt national energy security. Fracking is a burden Denton must shoulder for our country!

8. Hans Jonas, "Philosophical Reflections on Experimenting with Human Subjects," *Daedalus* 98, no. 2 (Spring 1969): 231.

9. See Gregg Easterbrook, *The Progress Paradox: How Life Gets Better While People Feel Worse* (New York: Random House, 2004).

10. Vaclav Smil, "World History and Energy," *Encyclopedia of Energy* 6:549–61, http://vaclavsmil.com/wp-content/uploads/docs/smil-article-2004world-history-energy.pdf.

11. Albert Borgmann, "The Nature of Reality and the Reality of Nature," in *Reinventing Nature? Responses to Postmodern Deconstruction*, ed. Michael E. Soule and Gary Lease (Washington, DC: Island Press), 31–45.

12. Marice Richter, "Exxon Mobile CEO Welcomes Fracking, but Not Water Tower in His Backyard," Reuters, February 26, 2014, http://www.reuters.com/article/2014/02/26/us-usa-fracking-tillerson-idUSBREA1P24O20140226.

13. John Locke, *Two Treatises of Government* (1689). The quote is from chapter 8 in the second treatise.

14. See Katinka Waelbers and Adam Briggle, "Three Schools of Thought on Freedom in Liberal, Technological Societies," *Techné* 14, no. 3 (Fall 2010): 176–93.

15. Robert McGinn, "Technology, Demography, and the Anachronism of Traditional Rights," *Journal of Applied Philosophy* 11, no. 1 (1994): 57–70.

16. Even F. A. Hayek, the great champion of individualism, knew this. See F. A. Hayek, "Individualism: True and False," in *Individualism and Economic Order* (Chicago: University of Chicago Press, 1948), 1–32.

17. This point seems true to being a proactionary. Precautionaries try to comprehend the whole and plan the big picture (often an angst-inducing way of being in the world). Proactionaries recognize our limited perspectives and focus on how individual contributions aggregate into something more (and something more unpredictable) than the sum of the parts.

18. Regarding the range of likely harms, see Lisa M. McKenzie, Roxana Z. Witter, Lee S. Newman, and John L. Adgate, "Human Health Risk Assessment

of Air Emissions from Development of Unconventional Natural Gas Resources," *Science of the Total Environment*, May 1, 2012, doi:10.1016/j.scitotenv.2012.02.018; and Peter Rabinowitz et al., "Proximity to Natural Gas Wells and Reported Health Status: Results of a Household Survey in Washington County, Pennsylvania," *Environmental Health Perspectives*, advance publication September 10, 2014, http://ehp.niehs.nih.gov/wp-content/uploads/advpub/2014/9/ehp.1307732.pdf. "Proximity" would have to be worked out also in terms of how people are tied together and what's most at risk. For example, if an entire parish in Louisiana relies on a single aquifer, then someone miles away from a proposed frack site arguably should have something close to the same say as someone right next to it.

19. See Elinor Ostrom, "Beyond Markets and States: Polycentric Governance of Complex Economic Systems," *American Economic Review* 100, no. 3 (June 2010): 641–72, http://bnp.binghamton.edu/wp-content/uploads/2011/06/Ostrom-2010-Polycentric-Governance.pdf.

20. Langdon Winner, *The Whale and the Reactor* (Chicago: University of Chicago Press, 1986).

21. Friedrich Engels, "On Authority" (1874), in *Marx-Engels Reader*, 2nd ed. (New York: Norton, 1978), 730–33.

22. Robert Nozick, *Anarchy, State, and Utopia* (New York: Basic Books, 1974), 268, 269.

23. Ibid., 334–35.

Chapter 7. HAND IN THE COOKIE JAR

1. Pam's video is linked in Peggy Heinkel-Wolfe, "Home Video, Personal Story of Friday's Gas Leak," *Denton Record-Chronicle*, April 24, 2013, http://newsgatheringblog.dentonrc.com/2013/04/home-video-personal-story-of-fridays-gas-leak.html/.

2. Leviticus 18:28.

3. Charles Perrow, *Normal Accidents: Living with High Risk Technologies* (Princeton: Princeton University Press, 1999).

4. Francis Bacon, *The New Organon* (1620), book 1, aphorism 3.

5. Ben Elgin, Benjamin Haas, and Phil Kuntz, "Fracking Secrets by Thousands Keep U.S. Clueless on Wells," Bloomberg, November 29, 2012, http://www.bloomberg.com/news/2012-11-30/frack-secrets-by-thousands-keep-u-s-clueless-on-wells.html.

6. There are other stories about slow response times and minimal action taken

when it comes to blowouts and other fracking accidents. For one example, see Deborah Sontag and Robert Gebeloff, "The Downside of the Boom," *New York Times*, November 22, 2014, http://www.nytimes.com/inter active/2014/11/23/us/north-dakota-oil-boom-downside.html?_r=0.

7. Peggy Heinkel-Wolfe, "Few Answers in April Gas Well Blowout," *Denton Record-Chronicle*, July 27, 2013, http://www.dentonrc.com/local-news/ local-news-headlines/20130727-few-answers-in-april-gas-well-blowout.ece.

8. The TCEQ report is available here: http://newsgatheringblog.dentonrc .com/files/2013/08/2240_001.pdf.

9. Rachael Rawlins, "Planning for Fracking on the Barnett Shale," *Virginia Environmental Law Journal* 31 (2014): 226–306.

10. Rachel Carson, *Silent Spring* (1962; Boston: Houghton Mifflin, 2002), 13.

11. See Michelle Bamberger and Robert Oswald, *The Real Cost of Fracking: How America's Shale Gas Boom Is Threatening Our Families, Pets, and Food* (Boston: Beacon Press, 2014). See also my review of this book, "Sentinels and Skeptics of the Shale," *Slate*, September 15, 2014, http://www.slate.com/ articles/technology/future_tense/2014/09/the_real_costs_of_fracking_a new_book_raises_questions_about_approaching.html.

12. For example, "Not only is it very difficult for residents to know whether a nearby operation poses risks to their health and families, and why, but regulators themselves are not capable of reliably answering that question." See Earthworks, *Blackout in the Gas Patch*, August 2014, http://www .earthworksaction.org/files/publications/BlackoutSummaryFINAL.pdf . And more generally, "The monitoring . . . has not kept up with the pace of development." See Sandy Bauers, "Greenspace: The Uncertain State of Gas Drilling and Health," *Philly.com*, March 2, 2014, http://www.philly.com/ philly/columnists/sandy_bauers/20140302_GreenSpace_The_uncertain_ state_of_gas_drilling_and_health.html.

13. Wolf Eagle Environmental, "Town of Dish, Texas Ambient Air Monitoring Analysis," 2009, http://townofdish.com/objects/DISH_-_final_report_ revised.pdf.

14. I can't say enough good things about this organization (www.shaletest.org), and I was thrilled and honored when they named Cathy, Sharon, and me as the recipients of the 2014 ShaleTest Superhero award!

15. TXsharon, "Property Devalued 75%!" *Daily Kos*, September 19, 2010, http://www.dailykos.com/story/2010/09/19/903261/-Property-devalued-75-Where-s-your-Drill-Baby-drill-now#.

16. Saul Elbein, "Here's the Drill," *Texas Monthly*, October 2011, http://www .texasmonthly.com/story/here's-drill?fullpage=1.

17. Wolf Eagle Environmental, "Mr. and Mrs. Timothy Ruggiero, 415 Star Shell Rd., Decatur, TX, 76234," http://www.earthworksaction.org/files/pubs-others/WolfEagleEnvironmentalReport.pdf.

18. Jim Efstathiou and Mark Drajem, "Drillers Silence Fracking Claims with Sealed Settlements," June 5, 2013, http://www.bloomberg.com/news/2013-06-06/drillers-silence-fracking-claims-with-sealed-settlements.html.

19. Caitlin Dickson, "Can You Silence a Child? Inside the Hallowich Case," *Daily Beast*, September 1, 2013, http://www.thedailybeast.com/articles/2013/09/01/can-you-silence-a-child-inside-the-hallowich-case.html.

20. Rawlins, "Planning for Fracking on the Barnett Shale," 245.

21. Ibid., 237–42.

22. A. G. Bunch et al., "Evaluation of Impact of Shale Gas Operations in the Barnett Shale Region on Volatile Organic Compounds in Air and Potential Human Health Risks," *Science of the Total Environment* 468–69 (2014): 832–42, http://www.sciencedirect.com/science/article/pii/S00489697130 10073.

23. Pamela Percival, "TCEQ Concerned About Emissions from Barnett Shale Facilities, Asks Industry to Help Find and Fix Emissions Problems," *Basin Oil and Gas*, February 2010, 16–23, http://web.archive.org/web/20120 403041758/http://www.fwbog.com/index.php?page=article&article=205.

24. TCEQ, Interoffice Memorandum from Shannon Ethridge, Toxicology Division, to Mark Vickery, Executive Director, May 25, 2010, http://www .tceq.state.tx.us/assets/public/implementation/barnett_shale/health Effects/2010.05.25-healthEffectsMemo.pdf.

25. See, for example, Gregg Macey et al., "Air Concentrations of Volatile Compounds Near Oil and Gas Production: A Community-based Exploratory Study," *Environmental Health* 13, no. 82 (2014), doi:10.1186/1476-069X-13-82.

26. On emissions inventories, see Rawlins, "Planning for Fracking on the Barnett Shale," section 2. Regarding trade-secret chemicals, one study shows data are available for only 14 percent of the products used by the industry, and for 43 percent of the products, less than 1 percent of total product composition is publicly available; see Theo Colburn et al., "Natural Gas Operations from a Public Health Perspective," *Human and Ecological Risk Assessment* 17, no. 5 (September 2011): 1039–56. On industry influence, see, for example, Public Citizen, "Drilling for Dollars," December 2010, http:// www.citizen.org/documents/drilling%20for%20dollars.pdf, which shows that incumbent Railroad Commission members get roughly 80 percent of their campaign contributions from the oil and gas industry. On not studying

impacts, see Katie Colaneri, "Former State Health Employees Say They Were Silenced on Drilling," NPR Pennsylvania State Impact Report, June 19, 2014, http://stateimpact.npr.org/pennsylvania/2014/06/19/former-state-health-employees-say-they-were-silenced-on-drilling/. See also David Hasemyer, "Two Texas Regulators Tried to Enforce the Rules. They Were Fired," *Daily Beast*, December 9, 2014, http://www.thedailybeast.com/articles/2014/12/09/two-texas-regulators-tried-to-enforce-the-rules-they-were-fired.html. On exemptions, see Renee Lewis Kosnik, "The Oil and Gas Industry's Exclusions and Exemptions to Major Environmental Statutes," Earthworks, October 2007, http://www.shalegas.energy.gov/resources/060 211_earthworks_petroleumexemptions.pdf.

27. Peggy Heinkel-Wolfe, "Residents Present Petition to City," *Denton Record-Chronicle*, May 7, 2014, http://www.dentonrc.com/local-news/local-news-headlines/20140507-petition-delivered-to-force-vote-on-fracking-ban.ece.

28. Jason Allen, "Is North Texas Gas Well Putting Cancer-Causing Chemicals in the Air?" CBS DFW News, May 1, 2014, http://dfw.cbslocal.com/2014/05/01/is-north-texas-gas-well-putting-cancer-causing-chemicals-in-the-air/.

29. See https://www.youtube.com/watch?v=GfxKY-nxBOU.

30. Allen, "Is North Texas Gas Well Putting Cancer-Causing Chemicals in the Air?"

31. Eduardo P. Olaguer, "The Potential Near-Source Ozone Impacts of Upstream Oil and Gas Industry Emissions," *Journal of the Air & Waste Management Association* 62, no. 8 (2012): 966–77, http://www.tandfonline.com/doi/pdf/10.1080/10962247.2012.688923.

32. Daniel Sarewitz, "How Science Makes Environmental Controversies Worse," *Environmental Science & Policy* 7 (2004): 385–403.

33. Adam Briggle, "Nature or Neoliberalism? Two Views on Science and the Persistence of Environmental Controversies," *Interdisciplinary Environmental Review* 15, nos. 2/3 (2014): 94–104.

34. Daniel Sarewitz, "Science and Environmental Policy: An Excess of Objectivity," in *Earth Matters: The Earth Sciences, Philosophy, and the Claims of Community*, ed. Robert Frodeman (Upper Saddle River, NJ: Prentice Hall, 2000), 79–98.

35. Sierra Crane-Murdoch, "Fallon, Nevada's Deadly Legacy," *High Country News*, March 9, 2014, https://www.hcn.org/issues/46.4/fallon-nevadas-deadly-legacy?src=me.

36. Ibid.

37. For example, Naomi Oreskes and Eric Conway, *Merchants of Doubt: How a Handful of Scientists Obscured the Truth on Issues from Tobacco Smoke to Global Warming* (New York: Bloomsbury, 2011).

38. See Rachel Aviv, "A Valuable Reputation," *The New Yorker*, February 10, 2014, 52–63.

39. Rose-Mary Sargent, "Philosophy of Science in the Public Interest: Useful Knowledge and the Common Good," Philosophy of Science Association 21st Biennial Meeting, Pittsburgh, PA, 2008, http://philsci-archive.pitt .edu/4300/.

40. The Meadows neighborhood is a case in point. Numerous families filed a similar trespass lawsuit against EagleRidge, but it went nowhere.

41. Aviv, "A Valuable Reputation," 58.

42. Rawlins, "Planning for Fracking on the Barnett Shale," 262, 264.

43. Jared Diamond, *Collapse: How Societies Choose to Fail or Succeed* (New York: Viking, 2005).

Chapter 8. MAKING RAIN

1. Earthworks, "Enforcement Report: RRC," September 2012, http://www .earthworksaction.org/issues/detail/inadequate_regulation_of_hydraulic_ fracturing#.VG5ZTqPnZD8.

2. Public Citizen, "Drilling for Dollars: How Big Money Has a Big Influence at the Texas Railroad Commission," December 2010, http://www.citizen.org/ drillingfordollars.

3. Kevin Roden, "Railroad Commission Chair Chimes in on Denton Frack Ban. And I Respond," July 12, 2014, http://rodenfordenton.com/2014/07/ railroad-commission-chair-chimes-in-on-denton-frack-ban-and-i-respond/.

4. Christian McPhate, "Water Woes: Drawn Away," *Denton Record-Chronicle*, June 8, 2014, http://www.dentonrc.com/local-news/local-news-headlines/ 20140608-water-woes-drawn-away.ece.

5. Brantley Hargrove, "The Effluent Society," *Texas Monthly*, June 2014, http:// www.texasmonthly.com/story/wichita-falls-drought-forces-town-to- drink-toilet-water.

6. Atul Gawande, "The Cancer-Cluster Myth," *The New Yorker*, February 8, 1999, http://www.newyorker.com/magazine/1999/02/08/the-cancer-cluster -myth.

7. See Ivan Illich, *H₂O and the Waters of Forgetfulness* (London: Marion Boyars, 1986).

8. See Christian McPhate, "New Report Looks at Fracking amid Drought," *Denton Record-Chronicle,* November 20, 2014, http://www.dentonrc.com/ local-news/local-news-headlines/20141120-new-report-looks-at-fracking-amid-drought.ece.

9. This is from a Montague County Property Owners Association newsletter and from a personal conversation in June 2014 with Robert McPhee, a member of that association. The newsletter is available here: http://www .themcpoa.com/documents/MCPOA%20News%20Letter%2009-30-13 .pdf.

10. McPhate, "Water Woes: Drawn Away."

11. Kate Galbraith, "Texas Study Finds Increase in Water Used for Fracking," *Texas Tribune,* January 15, 2013, https://www.texastribune.org/2013/01/15/ texas-study-traces-fracking-and-water-use/.

12. David Blackmon, "*Denton Record-Chronicle* Article Omits Key Facts About Water Use and Fracking," Energy in Depth, June 11, 2014, http://energyin depth.org/texas/denton-record-chronicle-omits-facts-water-fracking/.

13. Environmental Working Group, "Monster Wells," November 18, 2014, http://www.ewg.org/research/monster-wells.

14. Christian McPhate, "Krum, Vantage Energy Strike Deal for Water," *Denton Record-Chronicle*, October 13, 2014, http://www.dentonrc.com/local-news/ local-news-headlines/20141013-krum-vantage-energy-strike-deal-for-water .ece.

15. Michael Casey, "Drinking or Fracking," *Fortune*, September 2, 2014, http:// fortune.com/2014/09/02/fracking-drinking-water/.

16. McPhate, "Water Woes: Drawn Away."

17. Only 5 percent of fracking water is reused on the Barnett Shale. It's basically zero on the Eagle Ford Shale. See McPhate, "New Report Looks at Fracking amid Drought." But efforts are under way to increase the amount of reuse. See Nichola Groom, "Fracking Water's Dirty Little Secret—Recycling," Reuters, July 15, 2013, http://www.reuters.com/article/2013/07/15/ us-fracking-water-analysis-idUSBRE96E0ML20130715.

18. Tom Wilber, *Under the Surface: Fracking, Fortunes, and the Fate of the Marcellus Shale* (Ithaca: Cornell University Press, 2012). In Pennsylvania, shortly after treatment plants started accepting produced water, nearby mill workers noticed that the river water was corroding their equipment. The state put the Monongahela on a list of "impaired rivers," as 325,000

area residents were at one point told to drink bottled water. As an indication of the state of monitoring on the Marcellus, a 2013 study attempted to trace fracking water from withdrawal to disposal, but one of the scientists involved cited a lack of data and said, "We just couldn't do it." See Roger Real Druin, "As Fracking Booms, Growing Concerns About Wastewater," *Yale Environment 360*, February 18, 2014, http://e360.yale.edu/feature/ as_fracking_booms_growing_concerns_about_wastewater/2740/.

19. The information about updating comes from talking with Naomi Meier, who wrote her PhD dissertation on this topic.

20. Peggy Heinkel-Wolfe, "DA Office Dismisses Dumping Charges," *Denton Record-Chronicle*, October 28, 2012, http://www.dentonrc.com/local-news/ local-news-headlines/20121028-da-office-dismisses-dumping-charges.ece.

21. On stopping studies, see Abrahm Lustgarten, "EPA's Abandoned Wyoming Fracking Study One Retreat of Many," ProPublica, July 3, 2013, http://www .propublica.org/article/epas-abandoned-wyoming-fracking-study-one-retreat-of-many. For the Halliburton Loophole, see Environmental Integrity Project, *Fracking's Toxic Loophole*, October 2014, http://environmental integrity.org/wp-content/uploads/FRACKINGS-TOXIC-LOOPHOLE .pdf.

22. Steve Everley, "How Anti-fracking Activists Deny Science: Water Contamination," Energy in Depth, August 13, 2013, http://energyindepth.org/ national/how-anti-fracking-activists-deny-science-water-contamination/.

23. Numerous industry reports confirm the cracking and breaking. A good collection is Josh Fox, "The Sky Is Pink: Annotated Documents," http://www1 .rollingstone.com/extras/theskyispink_annotdoc-gasl4final.pdf. Groundwater contamination can occur not just from a fault or an old well creating a direct pathway from the shale. It's also caused by well-casing integrity and cementing failures. That explains why Pennsylvania documented water contamination from fracking: Lindsay Abrams, "Revealed: The 243 Times Fracking Contaminated Drinking Water in Pennsylvania," *Salon*, August 29, 2014, http://www.salon.com/2014/08/29/revealed_the_243_ times_fracking_contaminated_drinking_water_in_pennsylvania/. Even if hydraulic fracturing does not cause well-casing or cementing problems, it magnifies the negative consequences of those failures, because it introduces the toxic chemicals into the equation. So it would be correct to say that hydraulic fracturing polluted groundwater, because it was the source of the pollutants (even if not the cause of the migratory pathway). Further, there is remaining uncertainty about the role the increased pressures used in hydrau-

lic fracturing might play in causing well-casing failures. See Thomas Darrah et al., "Noble Gases Identify the Mechanisms of Fugitive Gas Contamination in Drinking-water Wells Overlying the Marcellus and Barnett Shales," *Proceedings of the National Academy of Sciences* 111, no. 39 (September 30, 2014): 14076–81, http://www.pnas.org/content/111/39/14076.full.

24. I had in mind the Lipsky case, where there was compelling evidence of contamination but the Railroad Commission continued to claim that "the evidence is insufficient" to draw any causal links. See WFAA Staff, "Scientists: Tests Prove Fracking to Blame for Flaming Parker County Wells," WFAA-ABC, June 5, 2014, http://www.wfaa.com/news/invest igates/Scientists-say-state-tests-prove-fracking-to-blame-for-Parker-Co-flaming-wells-262056131.html. See also Peter Gorman, "Proof," *Fort Worth Weekly*, June 25, 2014, http://www.fwweekly.com/2014/06/25/proof/.

25. Brian Fontenot et al., "An Evaluation of Water Quality in Private Drinking Water Wells Near Natural Gas Extraction Sites in the Barnett Shale Formation," *Environmental Science and Technology* 47, no. 17 (2013): 10032–40.

26. This is from an estimate based on how much waste it was permitted to handle. The figure is likely to keep climbing, as some estimate another 11,000 wells on the Barnett Shale are yet to be drilled: Gene Lockard, "Barnett Shale: Down, but Not Out," *Rigzone*, September 25, 2013, http://www.rigzone.com/news/oil_gas/a/129239/Barnett_Shale_Down_But_Not_Out.

27. Abrahm Lustgarten, "Injection Wells: The Poison Beneath Us," ProPublica, June 21, 2012, http://www.propublica.org/article/injection-wells-the-poison-beneath-us.

28. Dan Steward, *The Barnett Shale Play: Phoenix of the Fort Worth Basin, a History* (Fort Worth: Fort Worth Geological Society Press, 2007), 186.

29. Lustgarten, "Injection Wells."

30. Dave Fehling, "Silencing Those Who Would Scrutinize Disposal of Drilling Wastewater," NPR Texas State Impact Report, May 27, 2014, https://stateimpact.npr.org/texas/2014/05/27/silencing-those-who-would-scrutinize-disposal-of-drilling-wastewater/.

31. Lustgarten, "Injection Wells."

32. Mike Gaworecki, "Confirmed: California Aquifers Contaminated with Billions of Gallons of Fracking Wastewater," DeSmogBlog, October 7, 2014, http://www.desmogblog.com/2014/10/07/central-california-aquifers-contaminated-billions-gallons-fracking-wastewater.

33. Abrahm Lustgarten, "The Trillion-Gallon Loophole: Lax Rules for Drillers That Inject Pollutants into the Earth," ProPublica, September 20, 2012, http://www.propublica.org/article/trillion-gallon-loophole-lax-rules-for-drillers-that-inject-pollutants.

34. Michael Brick, "Vexed by Earthquakes, Texas Calls in a Scientist," *Houston Chronicle*, September 27, 2014, http://www.houstonchronicle.com/business/energy/article/Vexed-by-earthquakes-Texas-calls-in-a-scientist-5785791.php#/0.

35. Emily Atkin, "Texas Proposes Tougher Rules on Fracking Wastewater After Earthquakes Surge," Think Progress, August 27, 2014, http://thinkprogress.org/climate/2014/08/27/3476207/texas-earthquake-rules-fracking/.

36. Peggy Heinkel-Wolfe, "Petition Seeks Fracking Support," *Denton Record-Chronicle*, July 3, 2014, http://www.dentonrc.com/local-news/local-news-headlines/20140703-petition-seeks-fracking-support.ece.

37. TXsharon, "Caught on Video: Petition Circulators Deceive Denton Residents," Bluedaze blog, July 13, 2014, http://www.texassharon.com/2014/07/13/caught-on-video-petition-circulators-deceive-denton-residents/.

38. Peggy Heinkel-Wolfe, "Issue May Pack City Hall," *Denton Record-Chronicle*, July 12, 2014, http://www.dentonrc.com/local-news/local-news-headlines/20140712-issue-may-pack-city-hall.ece.

39. Edward Brown, "Compressing Agony," *Fort Worth Weekly*, May 14, 2014, http://www.fwweekly.com/2014/05/14/compressing-agony/2/.

40. You can watch the whole public hearing here: http://denton-tx.granicus.com/MediaPlayer.php?view_id=3&clip_id=989.

41. Perryman Group, "The Adverse Impact of Banning Hydraulic Fracturing in the City of Denton on Business Activity and Tax Receipts in the City and State," June 2014, http://energyindepth.org/wp-content/uploads/2014/07/Perryman-Denton-Fracking-Ban-Impact.pdf.

42. Amber took a video of my talk as seen at the Civic Center. It is available here: https://www.youtube.com/watch?v=BVhMdsenkMg.

43. The report from Curtin & Heefner is available here: http://newsgatheringblog.dentonrc.com/files/2014/07/0018_001.pdf.

44. I made a video featuring this moment called "Denton Takes a Stand." It is available here: https://www.youtube.com/watch?v=P1OQOsAiX_Y.

45. Dalton Gregory, "Reflecting on the Petition Initiative to Ban Fracking in Denton," July 22, 2014, http://www.daltongregory.com/reflecting-on-the-petition-initiative-to-ban-fracking-in-denton/.

Chapter 9. DAVID VS. GODZILLA

1. Peggy Heinkel-Wolfe, "Voters to Decide Ban Issue," *Denton Record-Chronicle*, July 15, 2014, http://www.dentonrc.com/local-news/local-news-headlines/20140715-voters-to-decide-ban-issue.ece.

2. For example, Max Baker, "Denton Leaders Submit Fracking Ban to the Voters," *Fort Worth Star-Telegram*, July 16, 2014, http://www.star-telegram.com/2014/07/15/5973163/denton-fracking-ban-hearing-draws.html.

3. See Kevin Roden, "Fracking and Denton's Economy: A Quick Response to the Perryman Study," July 15, 2014, http://rodenfordenton.com/2014/07/fracking-and-dentons-economy-a-quick-response-to-the-perryman-study/ and "Could a Frack-Free Denton Result in an Economic Boom?" July 14, 2014, http://rodenfordenton.com/2014/07/an-unlikely-economic-analysis-of-a-denton-fracking-ban/.

4. For example, Brad Johnson, "Fact Check: Keystone XL Tar-Sands Pipeline Will Not Create Jobs," *Grist*, November 7, 2011, http://grist.org/oil/2011-11-06-fact-check-keystone-xl-tar-sands-pipeline-will-not-create-jobs/.

5. I wrote this analysis up in a September 27 blog post that accompanied our first full-page Sunday ad in the *DRC*: http://frackfreedenton.com/2014/09/grow-strong-and-healthy/.

6. Adam Briggle, "Fracked on Their Own Petard," Denton Drilling blog, July 19, 2014, http://dentondrilling.blogspot.com/2014/07/depantsing-perryman-report-about.html, and *Truthout*, August 11, 2014, http://www.truth-out.org/opinion/item/25429-fracked-on-their-own-petard-the-self-implosion-of-an-industry-on-the-ropes#.

7. Holly Hacker and Jenna Duncan, "State Audit Bad News for UNT," *Denton Record-Chronicle*, September 24, 2014, http://www.dentonrc.com/local-news/local-news-headlines/20140924-state-audit-bad-news-for-unt.ece.

8. Georg Wilhelm Friedrich Hegel, *Philosophy of Right* (1820), preface.

9. In the science, technology, and society field, this is often called the Collingridge dilemma: Early in the process of technological development, there is too little information to anticipate problems. Later on, though, when we know about changes that should be made, it is too difficult to make them, because the technology is so entrenched (physically, economically, and politically). David Collingridge, *The Social Control of Technology* (New York: St. Martin's Press, 1980).

10. Karl Marx, *Theses on Feuerbach* (1845), XI.

11. Bill McKibben, "Bad News for Obama: Fracking May Be Worse than Burning Coal," *Mother Jones*, September 8, 2014, http://www.motherjones.com/environment/2014/09/methane-fracking-obama-climate-change-bill-mckibben. See also Christine Sherer et al., "The Effect of Natural Gas Supply on US Renewable Energy and CO_2 Emissions," *Environmental Research Letters* 9, no. 9 (2014), doi:10.1088/1748-9326/9/9/094008.

12. See G. Ottinger, "Changing Knowledge, Local Knowledge, and Knowledge Gaps: STS Insights into Procedural Justice," *Science, Technology & Human Values* 38, no. 2 (2013): 250–70.

13. Julie Dermansky, "Supporters of Fracking Ban Face New Wave of McCarthyism in Denton, TX," DeSmogBlog, October 22, 2014, http://www.desmogblog.com/2014/10/22/fracking-ban-supporters-face-new-wave-mccarthyism-denton-texas.

14. The law at issue here is Texas Local Government Code Chapter 245, "Issuance of Local Permits," http://www.statutes.legis.state.tx.us/Docs/LG/htm/LG.245.htm.

15. Joe Fisher, "History, Sprawl, Legacy Wells Put Fracking Focus on Texas Town," *NGI's Shale Daily*, April 10, 2014, http://www.naturalgasintel.com/articles/98018-history-sprawl-legacy-wells-put-fracking-focus-on-texas-town.

16. The vesting issue rubs up against the preemption issue as became apparent in the case of *Southern Crushed Concrete LLC v. City of Houston*. Notably, the law firm Baker Botts handled this case—the same law firm that would later represent the Texas Oil and Gas Association in their lawsuit against the Denton fracking ban.

17. Mr. Giovanetti would later characterize this as a form of tyranny. He'd likely make exceptions, though, in cases of clear threats to safety, such as in requirements to adopt new electrical or fire standards.

18. This question was at the heart of the influential 2011 case *City of Austin et al. v. Harper Park Two L.P.* You can find the decision of the Third Court of Appeals here: http://www.theaustinbulldog.org/index.php?option=com_docman&task=cat_view&gid=315&Itemid=22. To get a sense of the City of Denton's struggles with this, see Peggy Heinkel-Wolfe, "City's Position Paper on 'Vested Rights,'" *Denton Record-Chronicle* blog, November 24, 2014, http://newsgatheringblog.dentonrc.com/2014/11/citys-position-paper-on-vested-rights.html/.

19. Vesting is also an issue at the state level, as TCEQ often does not apply new air quality standards to existing gas well sites. Rachael Rawlins writes: "In

grandfathering existing facilities, the rulemaking leaves communities already suffering the impacts of existing gas operations without adequate protection." She quotes a TCEQ representative: "The most up-to-date science and emissions detection systems have greatly evolved over the past 25 years. Unfortunately, our laws have not." Rawlins concludes, "The TCEQ's decision to grandfather existing facilities burdens communities with industrial operations subject only to outdated and inadequate regulatory control." Rachael Rawlins, "Planning for Fracking on the Barnett Shale," *Virginia Environmental Law Journal* 31 (2014): 226–306, 291–93 quoted.

20. As my friend and former Denton City Council member Mike Cochran pointed out, English common law has long enshrined the concept of *sic utere tuo ut alienum non laedas* (do nothing with your land which will injure the land of another).

21. He posted them on his blog the next day: Tom Giovanetti, "Quotes from UNT Prof. Adam Briggle That Undermine His Call for a Ban on Fracking in the City of Denton, Texas," IPI Roundtable, September 3, 2014, http://www.ipi.org/policy_blog/detail/quotes-from-unt-prof-adam-briggle-that-undermine-his-call-for-a-ban-on-fracking-in-the-city-of-denton-texas.

22. Bj Lewis, "Area Republicans Oppose Fracking Ban," *Denton Record-Chronicle*, September 24, 2014, http://www.dentonrc.com/local-news/local-news-headlines/20140924-area-republicans-oppose-fracking-ban.ece.

23. I owe John Russell credit for this.

24. For the most part, we did a great job at this, but a few people admitted to me that they voted the wrong way.

25. They actually announced the decision in late August (the day after my meeting with the president), but it didn't get much play until September 4 and 5. See their press release here: http://business.denton-chamber.org/news/details/reasonable-regulation-urged-at-gas-well-drilling-sites.

26. Jenna Duncan, "Chamber Position Causing Problems," *Denton Record-Chronicle*, September 30, 2014, http://www.dentonrc.com/local-news/local-news-headlines/20140930-chamber-position-causing-problems.ece. Of course, it is ironic that our opponents would accuse us of being the group run by Washington, DC, interests!

27. Peggy Heinkel-Wolfe, "Chamber: Vote No to Fracking Ban," *Denton Record-Chronicle*, September 5, 2014, http://www.dentonrc.com/local-news/local-news-headlines/20140905-chamber-vote-no-to-fracking-ban.ece.

28. Kevin Roden, "We Need Solutions, Not Political Hackery," September 22, 2014, http://rodenfordenton.com/2014/09/we-need-solutions-not-political-hackery-a-response-to-the-chamber-inspired-fracking-mailing/.

29. Adam Briggle, "Shame on the Chamber Board," Denton Drilling blog, September 21, 2014, http://dentondrilling.blogspot.com/2014/09/shame-on-chamber-board.html.

30. Jaeah Lee and James West, "The Great Fracking Forward: Why the World Needs China to Frack Even More," *Wired*, September 18, 2014, http://www.wired.com/2014/09/great-fracking-forward/.

31. His story ran a week later: Clifford Kraus, "In Texas, a Fight Over Fracking," *New York Times*, October 9, 2014, http://www.nytimes.com/2014/10/09/business/in-texas-a-fight-over-fracking.html?_r=0.

32. I am transcribing this as best as I could from my notes—there were no recording devices allowed at this forum.

33. ShaleTest, "Project Playground," 2014, http://www.shaletest.org/wp-content/uploads/2014/09/ProjectPlaygoundPatagoniaReport-2.pdf.

34. Malcolm Gladwell, "The Unheard Story of David and Goliath," TED Salon NY2013, http://www.ted.com/talks/malcolm_gladwell_the_unheard_story_of_david_and_goliath?language=en.

Chapter 10. ELECTION DAY

1. By contrast, Denton Taxpayers for a Strong Economy didn't set up a Facebook page until late September and then quickly aborted it midsentence in their page description. It remained online, garnering a grand total of eleven Likes by Election Day. They did not do social media, and I wonder how much our big online, networked presence explains our eventual success. As Ed asked, "Do Likes translate into votes?" I still don't know the answer to that question.

2. Peggy Heinkel-Wolfe, "$500,000 Spent Trying to Turn Tide Just Before Vote," *Denton Record-Chronicle*, January 15, 2015, http://www.dentonrc.com/local-news/local-news-headlines/20150115-500000-spent-trying-to-turn-tide-just-before-vote.ece.

3. Frack Free Denton, "Fracking Carpetbaggers," October 8, 2014, http://frackfreedenton.com/2014/10/fracking-carpetbaggers/.

4. Denton Taxpayers for a Strong Economy, "Group Opposed to Denton Drill-

ing Ban Shows 98% of Contributions from Individuals and Businesses Paying Taxes in Denton," October 6, 2014, http://www.dentontaxpayers.com/press_release.

5. Denton Taxpayers for a Strong Economy, "Denton Drilling Ban Effort Tied to Russia," c. September 1, 2014, http://www.dentontaxpayers.com/denton _drilling_ban_effort_tied_to_russia.

6. Frack Free Denton, "Take Our Parks Back," October 15, 2014, http://frackfreedenton.com/2014/10/take-our-parks-back/.

7. Frack Free Denton, "Vote FOR the Children. Vote FOR the Ban," September 30, 2014, http://frackfreedenton.com/2014/09/vote-for-the-children/.

8. Terrence Stutz, "Texas Now 49th in Spending on Public Schools," *Dallas Morning News*, February 22, 2013, http://educationblog.dallasnews.com /2013/02/texas-now-49th-in-spending-on-public-schools.html/.

9. Peggy Heinkel-Wolfe, "Hot Under the Collar," *Denton Record-Chronicle*, October 16, 2014, http://www.dentonrc.com/local-news/local-news-head lines/20141016-hot-under-the-collar.ece.

10. This clash was a good reminder of the healthy differences of view on the DAG board. Another salient instance is differing opinions about the structure of organizations. Some favored more horizontal, egalitarian (even verging on leaderless) formats. Others favored stronger centralized control and clear chains of command. It's tempting to call this a generational divide, but I'm not sure that is the case.

11. Frack Free Denton, "Fracking Ban Opponents Attempt to Change Rules of Debate," October 10, 2014, http://frackfreedenton.com/2014/10/1379/.

12. Dianne Edmondson, "Act like Adults," *Denton Record-Chronicle*, October 14, 2014, http://www.dentonrc.com/opinion/letters-headlines/20141014-letters-to-the-editor-oct.-14.ece. See also "Frack Free Denton Afraid to Debate Ban?" on the Denton County Republican Party website, http://dentongop.org/uncategorized/frack-free-denton-afraid-debate-ban/.

13. I received the tweet within an hour of informing Ms. Edmondson about DAG's condition for participation. We did not know that Mr. Giovanetti had already been selected to represent the opposition.

14. Bj Lewis, "Residents Step Up to Debate," *Denton Record-Chronicle*, October 14, 2014, http://www.dentonrc.com/local-news/local-news-head lines/20141014-residents-step-up-to-debate.ece.

15. I had worked with Tara Linn's husband, Matthew, to make a great meme about this that said, "This isn't red or blue. It's black and white." It then

showed our black-and-white Frack Free Denton logo, saying, "Safer Neigh-borhoods. A Stronger Economy. That's something we can all get behind."

16. To see their performance, see Alisha Huber, "They're Living in a Disaster, so They Decided to Have a Little Fun," Upworthy, February 2, 2015. http://www.upworthy.com/theyre-living-in-a-disaster-so-they-decided-to-have-a-little-fun.

17. Michael Brick, "Fracking Ban Up for Vote in Denton, Texas," *Houston Chronicle*, November 3, 2014, http://powersource.post-gazette.com/power source/latest-oil-and-gas/2014/11/03/BC-DENTON-HNS/stories/201411030197.

18. Frack Free Denton, "Of Lawsuits and Lies," October 20, 2014, http://frackfreedenton.com/2014/10/of-lawsuits-and-lies/.

19. *Republic* 5.479e.

20. Dalton LaFerney, "Republicans to Blame for Polling Problems," *North Texas Daily*, October 30, 2014, http://ntdaily.com/republicans-to-blame-for-polling-problems/. After we won the election, the opposition put out an offensive (and wildly inaccurate) press release suggesting that college students are not real voters. See Jim Malewitz, "Industry Blames College Voters for Denton Fracking Ban," *Texas Tribune*, November 26, 2014, http://www.texastribune.org/2014/11/26/dentons-fracking-vote-numbers/.

Acknowledgments

1. Josh Harkinson, "Governor-Elect Laments the Californication of Texas," *Mother Jones*, January 12, 2015, http://www.motherjones.com/politics/2015/01/-greg-abbot-governor-californication-texas.

INDEX

Page numbers in *italics* refer to illustrations.